SPOUTED BEDS

SPOUTED BEDS

KISHAN B. MATHUR
and
NORMAN EPSTEIN

Department of Chemical Engineering
University of British Columbia
Vancouver, Canada

1974

ACADEMIC PRESS   New York   San Francisco   London

A Subsidiary of Harcourt Brace Jovanovich, Publishers

ACADEMIC PRESS, INC.
111 Fifth Avenue, New York, New York 10003

United Kingdom Edition published by
ACADEMIC PRESS, INC. (LONDON) LTD.
24/28 Oval Road, London NW1

Library of Congress Cataloging in Publication Data

Mathur, Kishan B
    Spouted beds.

    Bibliography: p.
    1.   Spouted bed processes.   I.   Epstein, Norman,
joint author.   II.   Title.
TP156.S57M37   660.2'84292   73-18975
ISBN 0–12–480050–5

Contents

v

13 Design

Appendix

References

Preface

When the idea of reviewing the field of spouted beds first occurred to us, in 1967, we both had no more than a sentimental interest in the subject, since we had separately been occupied with nonspouting activities for over a decade. The idea was prompted by the recurring appearance of publications on this topic which came to our notice from time to time. We embarked upon the reviewing task in 1968, visualizing a mini-monograph as the outcome of our effort. A little digging into the literature soon brought the realization that we had undertaken a considerably larger task than we had bargained for. As time went by, we also realized that we had re-entered a vigorously growing field and that spouting could not be simply characterized as "Canadian fluidization," to quote some of our colleagues from other countries. This view comes into sharp focus as we glance through the list of papers [273–286] which were presented in 1973 in Vancouver at the International Symposium on Spouted Beds.

The international character of work on spouted beds is also reflected in our main list of references, the language barrier having been overcome to a large extent through the help of many translators. Our Russian-to-English translator, Professor J. Solecki of UBC, had the busiest time, the contribution of Soviet workers to the field of spouted beds being very substantial.

We feel obliged, however, to take issue with Professor Zabrodsky concerning the origin of spouted beds. In the foreword of his book "Hydrodynamics and Heat Transfer in Fluidized Beds" (1966 [260]), Zabrodsky chides Leva (1959 [117]) for implying that spouted beds are ". . . a recent innovation hailing from Canada . . . ," and claims that ". . . we (in the Soviet Union) have known about air-fountain equipment for more than

20 years . . . ." On following up some of the references cited by Zabrodsky in support of this claim, we found that while the term "spouting" (literally "fountaining") in the Soviet technical literature indeed predates the Mathur–Gishler paper of 1955 [137], it was used by the Soviet workers in a different sense. For example, a system for burning crushed coal in which "the bed is strongly blown apart and fills the entire chamber" has been called a "spouting-type furnace" by Syromyatnikov (1951 [223]). The bulk density of coal as suspended in this system is given as 20–30 kg/m³ as against 400–500 kg/m³ for a packed bed of the same material. The operation described is not very different from that of a dilute-phase gas-fountain ore roaster patented in the USA almost a century ago by Robinson (1879 [196]), and bears little hydrodynamic resemblance to the well-defined dense-phase phenomenon termed "spouting" by Mathur and Gishler. Similarly, the "air-fountain" driers of Romankov *et al.* [200] and of Sazhin [207], also cited by Zabrodsky, are in fact dilute-phase systems, involving vigorous fluidization of dyes, pigments, and other paste-like materials in conical vessels. Thus, there appears to have been a semantic problem, which has since been resolved by Romankov and Rashkovskaya. These authors, in their book "Drying in a Suspended State" (1968 [201]), use the term "air-spouting" for the above type of dilute-phase operation, as distinct from "spouting bed" for the system which they note was "first investigated systematically" by Mathur and Gishler.

A large volume of information has since appeared in the technical literature on spouted beds, and some of it has been reviewed, from time to time, in journals [49, 8, 166, 60, 141] as well as books [117, 177, 169, 260, 201, 140]. Although the basic understanding of the spouting phenomenon is still far from complete, a stage has been reached where the accumulated information can be usefully compiled and an attempt made to consolidate what is known and identify what is not known about spouted beds. This is the primary objective of this book. The undertaking seems justified by the wide variety of processes for which spouting has already proved to be useful. It is hoped that the comprehensive state-of-the-art review which we have tried to present here will lead to fuller industrial exploitation of the special features of spouted beds on the one hand, and to further research activity into areas where better basic understanding is required on the other.

In keeping with the times, we have made a serious effort to use SI units, and the absence of g_c from the book bears testimony to the success of this effort. We have, however, taken certain liberties. Many of the empirical equations from the published literature have been left in their original units. We have felt more comfortable with hours and minutes than with kiloseconds. Centimeter has often been used, whenever the numbers happened to be more convenient in centimeters than in millimeters or kilometers. Temperatures are given in degrees Celsius rather than Kelvin, and angles are in degrees, not radians. Other deviations from the conventions of "strict SI" units also occur on occasion.

Acknowledgments

We are deeply grateful to workers in several countries who kept us supplied with not only reprints but also preprints of their publications, as well as with their books and dissertations, during the last five years. Notable among these were Professor Ratcliffe of Australia, Professors Kugo and Uemaki of Japan, Professor Németh and Dr. Pallai of Hungary, Professors Volpicelli and Massimilla of Italy, Professors Vuković and Zdanski of Yugoslavia, Professors Nelson and Clary of the USA, and Professors Mukhlenov, Gorshtein, Romankov, Zabrodsky, Mikhailik, Gelperin, and Ainshtein of the USSR. We are also indebted to Dr. Storrow (Fisons, England), Mr. Berquin (PEC, France), Mr. McDonnell (Cominco, Canada), and especially to Mr. Peterson (NRC, Canada), who all provided us with valuable unpublished information as cited in the text.

The undergraduate chemical engineering students at the University of British Columbia have made a significant contribution to the book, and this has been acknowledged in the list of references, which includes no less than a dozen UBC Bachelor's theses. Assistance with computational work was also provided by graduate students C. J. Lim and T. L. Truong.

We would like to pay a special tribute to Laurie Sharpe who, in addition to her normal duties, somehow managed to produce draft after draft of the manuscript on her typewriter with remarkable speed, meticulous care, and always in good cheer. Veena Mathur provided substantial support for the project in many significant ways.

Finally, we are grateful to the National Research Council of Canada for financial assistance, provided through operating research grants.

1 | *Introduction*

1.1 THE SPOUTED BED

Consider a vessel open at the top and filled with relatively coarse particulate solids. Suppose fluid is injected vertically through a centrally located small opening at the base of the vessel. If the fluid injection rate is high enough, the resulting high velocity jet causes a stream of particles to rise rapidly in a hollowed central core within the bed of solids. These particles, after reaching somewhat above the peripheral bed level, rain back onto the annular region between the hollowed core and the column wall, where they slowly travel downward and, to some extent, inward as a loosely packed bed. As the fluid travels upward, it flares out into the annulus. The overall bed thereby becomes a composite of a dilute phase central core with upward moving solids entrained by a cocurrent flow of fluid, and a dense phase annular region with countercurrent percolation of fluid. A systematic cyclic pattern of solids movement is thus established, giving rise to a unique hydrodynamic system which is more suitable for certain applications than more conventional fluid–solid configurations.

1

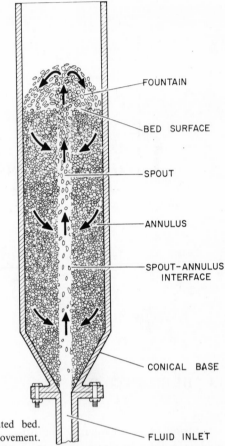

FOUNTAIN

BED SURFACE

SPOUT

ANNULUS

SPOUT–ANNULUS
INTERFACE

CONICAL BASE

FLUID INLET

Fig. 1.1. Schematic diagram of a spouted bed.
Arrows indicate direction of solids movement.

This system is termed a *spouted bed,* the central core is called a *spout,* and the peripheral annular region is referred to as the *annulus.* The term *fountain* will be used to denote the mushroom-shaped zone above the level of the annulus. To enhance solids motion and eliminate dead spaces at the bottom of the vessel, it is common to use a diverging conical base, with fluid injection at the truncated apex of the cone (see Fig. 1.1). The vessel itself is usually a circular cylinder, though the use of an entirely conical vessel has been common practice in the USSR. An example of each is shown in Fig. 1.2. Since spouting in conical vessels has been extensively discussed in the recent book by Romankov and Rashkovskaya [201], the emphasis of the present book is on spouting in cylindrical columns.

Solids can be added into and withdrawn from a spouted bed, so that, like fluidization, spouting lends itself to continuous operation. The solids may be fed into the bed either at the top near the wall, so that they join the

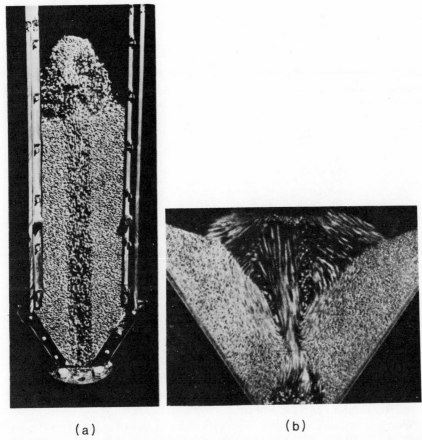

Fig. 1.2. (a) Spouting in a cylindrical vessel [137]; (b) spouting in a conical vessel [201].

downward moving mass of particles in the annulus, or with the incoming gas (see Fig. 1.3). Since the annular solids are in an aerated state, material discharges readily through an overflow pipe located in the column wall or through an outlet at a lower level. When solids are fed with the inlet gas and discharged through an overflow pipe, as in Fig. 1.3b, they are effectively being conveyed vertically in addition to being contacted with the gas.

The spouted bed was originally conceived and has hitherto been regarded primarily as a modified version of a fluidized bed, the need for modification arising from the poor quality of fluidization encountered with uniformly coarse particles. Thus, some recent books on fluidization [242, 110] have characterized a spouted bed simply as a special type of fluidized bed. This view is no longer adequate, since developments over the last ten years have demonstrated that a spouted bed displays special characteristics which

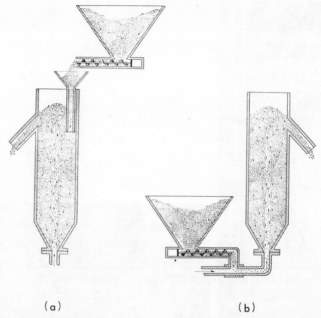

(a)   (b)

Fig. 1.3. Continuous spouting operation. Controlled discharge of solids can also be effected at lower levels. (a) Solids fed into annulus (after Mathur and Gishler [138]); (b) solids fed with incoming gas (after Manurung [134]).

render it capable of performing certain useful cyclic operations on solid particles which cannot be performed in a fluidized bed due to its comparatively random particle motion.

Spouting action can be obtained with either a gas or a liquid as the jet fluid. As a substitute for fluidization, liquid spouting does not offer the same advantage as gas spouting since the problem of fluidization quality, even with coarse particles, is generally eliminated in a liquid medium. The systematic cyclic movement of particles remains a distinctive feature of liquid spouting, but no attempts appear to have been made to turn this property to practical advantage. Work on spouting has, therefore, been overwhelmingly with gas media (usually air), and it will be assumed throughout this book that the jet fluid is a gas, unless otherwise indicated.

1.2 HISTORY

The terms *spouted bed* and *spouting* were coined at the National Research Council of Canada in 1954 by Gishler and Mathur [70]. These investigators developed this technique initially as a method for drying wheat. They were able to use much hotter air than in conventional wheat driers [138] without damaging the grain. Realizing that the technique could have wider application, they studied the characteristics of a spouted bed using a variety of solid materials with both air and water as the spouting medium [137]. On the basis of this preliminary study they were able to assert that "the mechanism of flow of solids as well as of gas in this technique is different from fluidization, but it appears to achieve the same purpose for coarse particles as fluidization does for fine materials."

Early research papers continued to originate from Canadian sources— the National Research Council Laboratories in Ottawa, the Prairie Regional Laboratories of the NRC, and the University of Ottawa. It was only after the publication in 1959 of Leva's book on fluidization [117], which summarized the NRC work in a chapter entitled "The Spouted Bed," that interest in the technique spread to other countries—if one takes publication as an index of interest. Nearly 200 publications, including patents, on various aspects or applications of spouted beds have since appeared from Australia, Canada, France, Hungary, India, Italy, Japan, Rumania, the United Kingdom, the USA, Yugoslavia and, most prolifically, the USSR.

The translation of Leva's book into Russian in 1961, followed by the publication in 1963 of a book by Zabrodsky [260] in which an excellent chapter on spouted beds summarized the early Canadian work, appears to have triggered considerable activity on the subject at several research centers in the Soviet Union. Particular emphasis has been placed on spouting in a conical vessel rather than in a cylindrical column with a conical base. The hydrodynamics of the system has received careful attention, and applications have been developed for a wide range of functions.

The first commercial spouted bed units in Canada were installed in 1962 —for drying peas, lentils, and flax. Units have since been built in several other countries for a variety of other drying duties, including evaporative crystallization, as well as for solids blending, cooling, coating, and granulation. The successful application of the spouted bed technique to the granulation of fertilizers and other products by Berquin [19, 20] is particularly worthy of note, since it represents an application which takes advantage of the unique solids recirculation feature of a spouted bed. Similar developments have also been made in the USSR.

Potential industrial applications of spouted beds which are as yet in the experimental stage include coal carbonization, shale pyrolysis, ore roasting, and even petroleum cracking. With the variety of applications already commercialized and an even wider variety in the offing, spouting appears to have come of age. It is no longer an esoteric method for drying grain but a full-fledged process engineering operation with a substantial body of technical literature, an increasing number of practical applications, and a developing technology.

1.3 REQUIREMENTS FOR SPOUTING

Spouting, which is a visually observable phenomenon, occurs over a definite range of gas velocity for a given combination of gas, solids, and vessel configuration. Figure 1.4 illustrates the transition from a quiescent to a spouting to a bubbling to a slugging bed, which often occurs as gas velocity is increased.

These transitions can be represented quantitatively as plots of bed depth versus gas velocity, or *phase diagrams,* examples of which are given in Figs. 1.5a–1.5d. The line representing transition between a static and an

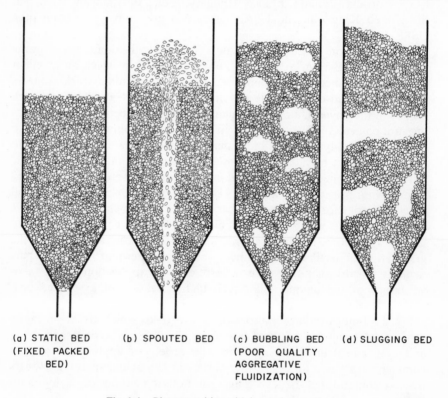

(a) STATIC BED
(FIXED PACKED
BED)

(b) SPOUTED BED

(c) BUBBLING BED
(POOR QUALITY
AGGREGATIVE
FLUIDIZATION)

(d) SLUGGING BED

Fig. 1.4. Phase transition with increasing gas flow.

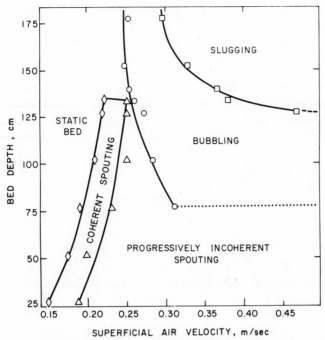

Fig. 1.5a. Phase diagram. Sand, $d_p = 0.42$–0.83 mm, $D_c = 15.2$ cm, $D_i = 1.25$ cm [137]. In progressively incoherent spouting, fountain shape becomes ill-defined, particle motion above bed gets increasingly chaotic, but annular solids movement remains intact.

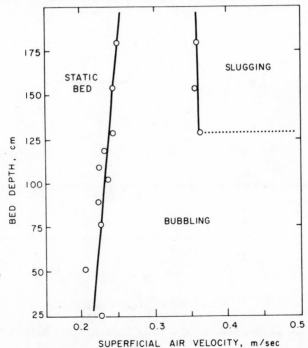

Fig. 1.5b. Phase diagram. Sand, $d_p = 0.42$–0.83 mm, $D_c = 15.2$ cm, $D_i = 1.58$ cm [137].

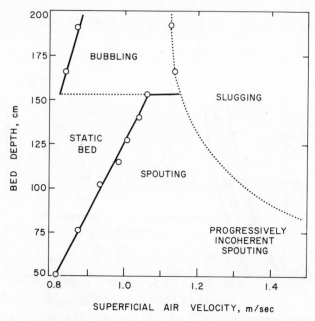

Fig. 1.5c. Phase diagram. Wheat, $d_p = 3.2 \times 6.4$ mm, $D_c = 15.2$ cm, $D_i = 1.25$ cm [137].

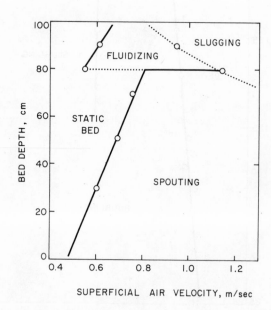

Fig. 1.5d. Phase diagram. Semicoke, $d_p = 1$ mm–5 mm, $D_c = 23.5$ cm, $D_i = 3.05$ cm [49].

agitated (spouted or fluidized) bed is more reproducible in the direction of decreasing velocity than vice versa, the resulting static bed being in the reproducible *random loose packed* condition [54]. Figures 1.5a–1.5d show that, for a given solid material contacted by a specific fluid in a vessel of fixed geometry, there exists a *maximum spoutable bed depth* H_m, beyond which spouting action does not occur but is replaced by poor quality fluidization. The *minimum spouting velocity* at this bed depth can be up to 50% greater than the corresponding minimum fluidization velocity U_{mf} [173], though closer correspondence between these two critical velocities has usually been found [15, 114] and for analytical purposes it has sometimes been convenient to equate them [228]. Phase diagrams of Figs. 1.5a and 1.5b also indicate that for given solids, gas, and column diameter, there is a maximum gas inlet size beyond which spouting does not occur, the bed changing directly from the quiescent to the aggregatively fluidized state.

More generalized phase diagrams have been attempted by Pallai and Németh [173], who plotted pressure drop versus reduced velocity U/U_{mf} for various bed depths, and by Becker [15], who used similar parameters, all normalized with respect to the condition of minimum spouting at H_m. Although there is some validity and much elegance in these normalizing procedures, the resulting generalizations oversimplify the transition behavior, which is governed by complex interrelationships between the variables involved.

A typical spouted bed has a substantial depth, which in the case of a cylindrical vessel is usually of the order of at least one column diameter, measured from the inlet orifice to the surface of the annulus. If the bed is much shallower, the system becomes hydrodynamically different from true spouting, and any generally formulated principles of spouted bed behavior would not be expected to apply. A *minimum spoutable depth* has, however, not been precisely defined or investigated. Nor have any detailed studies been made about the *maximum spouting velocity,* at which transition from coherent spouting to either bubbling fluidization or slugging occurs. For most practical purposes, however, there is usually ample latitude between the minimum and maximum spouting velocity so that the velocity can be increased by a large factor without transition to fluidization.

1.4 SPOUTED BEDS IN THE GAS–SOLIDS CONTACTING SPECTRUM

The more common gas–solids contacting systems may be broadly classified as (a) nonagitated, (b) mechanically agitated, and (c) gas agitated.

Fixed and moving packed beds, which fall in the first category, are applicable to processes which do not call for high rates of heat and mass transfer between the gas and the coarse solids, and in which uniformity of conditions in different parts of the bed is either not critical or not desirable. In a fixed bed, solids cannot be continuously added or withdrawn, and treatment of the gas is usually the main objective—for example, recovery of solvent vapors by adsorption, gas drying, and catalytic reactions with a catalyst of long life. A moving bed does provide continuous flow of solids through the reaction zone. Its application therefore extends to solids treatment in such uses as roasting of ores, calcining of limestone, and drying and cooling of pellets and briquettes. In both fixed and moving beds, the gas movement is close to *plug flow*, a feature which is advantageous for certain exothermic chemical reactions where it is desired to have the temperature increase as the reaction proceeds, that is, as the driving force diminishes.

Limited agitation can be imparted to the solids by mechanical means, either by movement of the vessel itself, as in rotary driers and kilns, or by the use of internal agitators. In either case, most of the material at any instant is still maintained in a packed bed condition, but the relative movement of particles improves contacting effectiveness since fresh surface is continually exposed for action by the gas. Also, the blending of solids by agitation levels out interparticle gradients in composition and temperature. Mechanical systems are mainly used for processes involving solids treatment, such as drying, calcining, and cooling, but are obviously unsuitable for processes which require the gas to be uniformly treated.

In gas-agitated systems such as fluidized and suspended beds, a more intense form of agitation is imparted to each solid particle by the action of the gas stream. In dense phase fluidization, the bed of fine solids behaves as a well-mixed body of liquid, and the high solids surface involved gives rise to high particle-to-gas heat and mass transfer rates. A stream of solids can be easily added to and withdrawn from the bed. Because of these basic features, the fluidized bed has emerged as the preferred contacting method for a number of processes which include chemical reactions (both catalytic and noncatalytic) as well as treatment of solids. In fluidization, the presence of a coherent bed of solids allows the solids flowing through the system to spend a certain time in the reaction zone, which can be controlled by adjusting the ratio between the feed rate and the weight of the bed. The bed also acts as a buffer to dampen out any instabilities which arise during continuous operation.

Dilute phase gas–solids contacting (suspended bed or transport system) may be more suitable than fluidization in other process situations. In such systems the contact time between a given particle and a gas is very short— no more than a few seconds—because of very high gas velocities. Intense turbulence makes for high coefficients of heat and mass transfer, but the extents of heat and mass transfer for the particles are not high because of their small residence time in the reaction zone. Dilute phase systems are suitable for processes in which the gas–solids interaction is surface-rate-controlled rather than diffusion-controlled. Examples of established applications are combustion of pulverized coal, flash roasting of metallic sulfides, and drying of sensitive materials which can tolerate exposure to heat for only a few seconds.

In the Chemical Engineers' Handbook [177] spouted beds are placed in the category of moving-bed systems, presumably because the annular region, which contains most of the solids, does constitute a moving bed with countercurrent flow of gas. However, inasmuch as the bed solids are well mixed, that is, recirculated many times, the operation comes closer to fluidization in its characteristics. While the original statement that "spouting appears to achieve the same purpose for coarse particles as fluidization does for fine materials" [137] remains true, it is now apparent that the systematic cyclic movement of particles in a spouted bed, as against the more random motion in fluidization, is a feature of critical value for certain applications, for instance, in granulation and particle coating processes. In the spectrum of contacting systems, then, spouted beds occupy a rather complex position, overlapping fluidized and moving beds to some extent, but at the same time having a place of their own by virtue of certain unique characteristics.

Apart from achieving gas–solids contact, spouting is also a means of agitating coarse particles with a gas and of causing solid–solid contacts to occur, the interaction between the gas and solid particles being incidental. These additional features of a spouted bed can be applied to perform mechanical operations such as blending, grinding, and dehusking of solids.

1.5 SPOUTING VERSUS FLUIDIZATION

The minimum particle diameter for which spouting appears to be practical is about 1 mm. This particle size is close to the value above which the gas–solids contacting effectiveness of fluidized beds is seriously impaired due to bypassing of gas in the form of large bubbles [46, p. 8; 113]. It is possible to operate a miniature spouted bed with considerably finer particles [75, 11, 1, 203] using a very small gas inlet. Indeed, use of sieve plate distributors for fluidized beds gives rise to spout formation above each hole, the spouts breaking into bubbles further up the bed [114, 61a]. If, however,

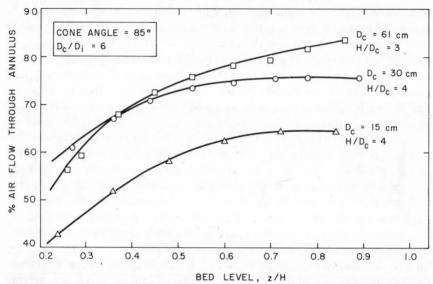

Fig. 1.6. Air distribution in spouted wheat beds [228]. Reproduced by permission of *The Canadian Journal of Chemical Engineering* **37**, 184 (1959), Fig. 4.

a single small hole is used to spout fine particles, the allowable holdup and capacity of the bed would also be small, and any attempt at scale-up by enlarging the inlet hole would give rise to nonhomogeneous fluidization rather than spouting, even for a conical vessel [71, 241].

The total frictional pressure drop ΔP_s across a fully spouting bed is always lower by at least 20% [166] than that required to support the weight of the bed, that is, than the frictional pressure drop for particulate or good quality aggregative fluidization. In this respect, a spouted bed is qualitatively similar to a channeling fluidized bed [263, p. 22]. Channeling, however, is an undesirable feature of a fluidized bed, involving the passage of gas through part of the bed without inducing much movement in the surrounding particles. In spouting, on the other hand, agitation of the entire bed is achieved by means of the gas jet, and intimate contact between the particles and the gas occurs both in the jet and in the dense phase region, the latter being fed by cross flow from the former. This is illustrated in Fig. 1.6, which shows the large proportion of gas which actually permeates the annulus by the time it reaches the upper half of the bed. In addition, channeling in fluidized beds is favored by very fine particles [136; 116; 117, p. 24; 263, p. 261; 6], while spouting is normally a coarse particle phenomenon. Thus the similarity between spouting and fluidized bed channeling emphasized by several authors [117, p. 170; 260, p. 111; 46, p. 6], is somewhat misleading.

The shape of the longitudinal pressure profile in a spouted bed is also distinctly different from that of a fluidized bed, even when a conical-cylindrical vessel with no bed support is used in both cases. Thus for a spouted bed the longitudinal pressure gradient varies with bed level, rising steadily to a maximum value at the top of the bed. For a fluidized bed, on the other hand, the pressure gradient is constant over the cylindrical portion of the column, despite the fact that the general pattern of solids movement in this case also is up at the center and down near the wall.

1.6 LAYOUT OF SUBJECT MATTER

Chapters 2–6 deal with the fluid and solid dynamics of spouted beds. Chapter 2 is concerned with the mechanism of transition from a packed bed to a spouted bed, and presents methods for predicting spouting velocity and pressure drop requirements. Chapter 3 details the flow pattern of gas in both annulus and spout, and Chapter 4 that of solids, including recirculation and gross mixing behavior. This is followed by a chapter on the internal geometrical structure of a stable bed. Spouting stability, with particular reference to the estimation of maximum spoutable bed depth, is discussed in Chapter 6.

Chapter 7 is devoted to the subject of attrition, often a shortcoming of spouted beds, which has, however, been turned to advantage in several applications.

Chapters 8 and 9, respectively, examine heat and mass transfer in spouted beds, both between fluid and particles and within particles. Chapter 8 also covers heat transfer between the column wall and the bed, as well as between a submerged object and the bed, while the mass transfer discussion focuses on solids drying. Next, the question of using a spouted bed for carrying out gas phase chemical reaction is considered in Chapter 10, the only entirely theoretical chapter in the book.

Chapter 11 describes the large variety of mechanical, thermal, diffusional, and chemical processes to which spouted beds have been applied, whether on the bench, pilot, or commercial scale. Also included are suggestions for potential applications of spouted beds. This is followed by a review of the various process and equipment modifications to a "standard" spouted bed which have been devised to achieve specific ends.

Finally, Chapter 13 applies much of the previous material to roughing out a rudimentary design strategy, and includes some practical hints for the benefit of the spouted bed designer and operator.

2 | *The Onset of Spouting*

2.1 MECHANISM

The mechanism of transition from a static to a spouted bed is best described with reference to a plot of bed pressure drop versus superficial velocity. The experimental data of Lama and co-workers [111, 125] are shown in Fig. 2.1. The plot for the deepest bed (ABCD) is supplemented by a dashed curve (DC′B′A) illustrating the reverse process, that is, the collapse of a spouted bed on decreasing the gas velocity, based on the work of Mathur, Thorley, and co-workers [137, 227, 228]. For observing the corresponding physical behavior, all the above workers employed half-round sectional columns in which the axial zone of the bed could be seen against the flat transparent face.

The following sequence of events is observed as the gas flow is increased (see Fig. 2.1, top curve):

(1) At low flow rates the gas simply passes up without disturbing the particles, the pressure drop rising with flow rate (along AB) as in any static packed bed.

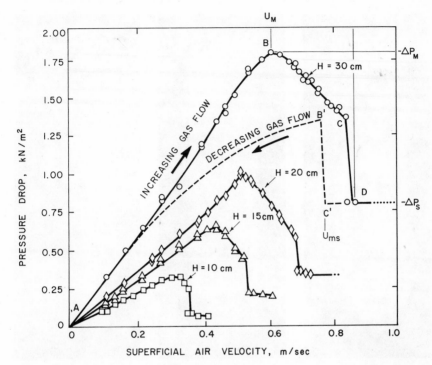

Fig. 2.1. Typical pressure drop–flow rate curves (after Madonna *et al.* [125]). Wheat: $d_p = 3.6$ mm, $D_c = 15.2$ cm, $D_i = 1.27$ cm, $\theta = 60$.

(2) At a certain flow rate, the jet velocity becomes sufficiently high to push back the particles in the immediate vicinity of the gas inlet, causing a relatively empty cavity to form just above the inlet. The particles surrounding the cavity are compressed against the material above, forming a compacted arch which offers a greater resistance to flow (see Fig. 2.2a). Therefore, despite the existence of a hollow cavity, the total pressure drop across the bed continues to rise.

(3) With further increase in gas flow, the cavity elongates to an internal spout (Fig. 2.2b). The arch of compacted solids still exists above the internal spout so that the pressure drop across the bed rises further until it reaches a maximum value $-\Delta P_M$, at point B. The corresponding superficial velocity is denoted by U_M.

(4) As the flow rate is increased beyond point B, the height of the relatively hollow internal spout becomes large in comparison with the packed solids above the spout. The pressure drop therefore decreases along BC.

(5) As point C is approached, enough solids have been displaced from the central core to cause a noticeable expansion of the bed. This bed ex-

                    (a)   (b)   (c)

Fig. 2.2. The onset of spouting: (a) Small cavity forms; (b) internal spout develops; (c) steady spouting sets in. Courtesy of W. S. Peterson, NRC, Ottawa.

pansion sometimes results in arresting the fall in pressure drop [228] (though there is little such tendency exhibited in the curves of Fig. 2.1), and is usually accompanied by alternate expansion and contraction of the internal spout. The resulting instability gives rise to pressure drop fluctuations [201, p. 42] and for deeper beds, to fluidization of the particles above the internal spout [173].

(6) With only a slight increase in flow rate beyond point C, which is called the point of *incipient spouting*, the internal spout breaks through the bed surface (Fig. 2.2c). When this happens, the solids concentration in the region directly above the internal spout decreases abruptly, causing a sharp reduction in pressure drop to point D, at which the entire bed becomes mobile and steady spouting sets in. Point D thus represents the *onset of spouting*.*

   *In the early papers on spouted beds, D was usually labeled the incipient spouting point, but subsequently this designation was reserved for the more unstable condition corresponding to point C.

(7) With still further increase in gas flow, the additional gas simply passes through the spout region, which is now established as the path of least resistance, causing the fountain to shoot up higher without any significant effect on the total pressure drop. The pressure drop $-\Delta P_s$ beyond point D therefore remains substantially constant.

The incipient spouting velocity (C) and the onset of spouting (D), being bed-history dependent, are not exactly reproducible. A more reproducible velocity, the *minimum spouting velocity* U_{ms}, is obtained by slowly decreasing the gas flow: the bed then remains in the spouted state until point C', which represents the minimum spouting condition. A slight reduction of gas velocity at this condition causes the spout to collapse and the pressure drop to rise suddenly to point B'. Further diminution of flow rate causes pressure drop to decrease steadily along B'A. However, the main curve now falls lower than for increasing flow since the energy required by the gas jet to

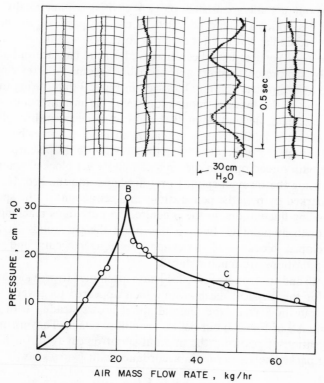

Fig. 2.3. Instantaneous and average static pressures measured at $z = 4$ cm on the nozzle axis [250]. Glass spheres, $d_p = 3.1$ mm, $H = 22$ cm.

penetrate the solids is no longer expended during the collapse of the spout. The hysteresis loop ABCDC'B'A can be shrunk somewhat if the process is repeated starting with a loose rather than a relatively dense packed bed, but it cannot be eliminated entirely due to the inherent irreversibility of the jet penetration phenomenon.

A more detailed study of the transition from a quiescent to a spouting bed has been reported by Volpicelli and Raso [250], who carried out instantaneous pressure measurements with an electronic transducer near the base of the bed. Their column was rectangular in cross section, 200 × 15.5 mm, with the gas entering through a centrally located 4 mm wide slot running the 15.5 mm between the column faces. The development of the internal spout and the particle trajectories were recorded with a motion picture camera. Plots of instantaneous and time-average pressure published by these workers are reproduced in Fig. 2.3. The latter is labeled to match the upper curve of Fig. 2.1.

Volpicelli and Raso observed that in their two-dimensional column, the internal spout actually consisted of two cavities symmetrically located about the gas orifice axis and bounded by moving particles, the flow lines of which were similar to those for a pair of vortices in a continuous medium. Similar pairs of vortices have been observed on the flat face of half-round columns by the present authors and probably become a vortex ring in the case of an axisymmetric column. It is seen in Fig. 2.3 that during the initial development of the internal spout along AB, the instantaneous pressure does not fluctuate. Beyond point B the gas jet penetrates through a zone of low resistance in the compacted arch, forming a somewhat winding channel through the bed and causing the pressure to drop sharply. At this stage, the path forced open in the arch can easily get blocked again by the solids, causing the through-spout to retract back into the bed. Hence, this state is marked by periodic penetrations and retractions of the spout with corresponding fluctuations in the pressure, the pulsations reaching a maximum growth at point C. Beyond this point the gas–solids jet becomes fully developed, blockage of the channel no longer occurs, and therefore pressure pulsations are much reduced.

Volpicelli and co-workers [250, 251] have attempted to provide theoretical support for the physical mechanism which appeared to govern the early stage of transition, when the internal spout grows steadily with gas flow rate (along AB of Fig. 2.3) but before instability sets in. They assumed that the permeation of gas from the internal spout or cavity through the surrounding packed solids occurs in accordance with Darcy's law,

$$\nabla P = KU \qquad\qquad (2.1)$$

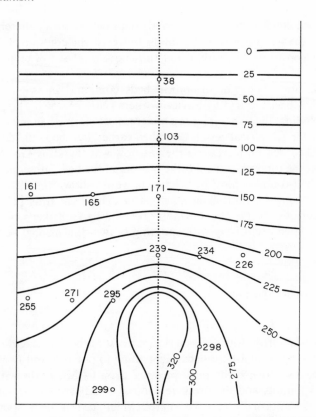

Fig. 2.4. Pressure distribution during transition: Observed (O) and calculated (_____) values, mm water [251].

Neglecting any density variation of the gas and assuming isotropy of the packed solids, it follows that

$$\nabla^2 P = 0 \tag{2.2}$$

The above Laplace equation was numerically integrated by a finite difference method in the domain of the two-dimensional packed solids. The two boundary conditions were the measured constant cavity pressure along the observed contour of the cavity, and atmospheric pressure at the upper surface of the bed. The computed pressure distribution is compared with a number of measured values in Fig. 2.4. It is seen that the agreement is fairly good, the small discrepancies probably being due mainly to compaction of solids in the arch immediately surrounding the cavity, mentioned in (2) (p. 15). This agreement confirms that, over the stable part of the transition, the bed solids outside the internal spout behave as a coherent packed bed. Once the gas jet pierces the solids, however, the contour of the cavity no longer constitutes an isobar and the above analysis, of course, breaks down.

The transition behavior of conical beds contained in two-dimensional columns has been studied by Soviet workers [201, 79, 256]. Their observations, including the pressure drop diagram, conformed in most respects with those for cylindrical vessels described earlier, but the initial disturbance of the packed bed on increasing the gas flow was found by Goltsiker [79] to occur at the top rather than at the bottom of the bed. Subsequent experiments in half-round conical beds have, however, shown the disturbance to propagate from the bottom upward, as in cylindrical beds [12, 257], so that the curious reverse phenomenon observed by Goltsiker would appear to be associated exclusively with spouting in two-dimensional beds.

2.2 PRESSURE DROP

The pressure drop values of practical interest in the design and operation of a spouted bed unit are those corresponding to points B and D in Fig. 2.1. These are, respectively, the peak pressure drop $(-\Delta P_M)$ attained prior to incipient spouting, and the pressure drop at steady spouting $(-\Delta P_s)$. The former would normally be encountered when starting up a spouted unit (unless an axial draft tube is used—see Section 11.4), while the latter would determine the operating power requirement.

A. Interpretation of Pressure Drop Measurements

Since spouted bed pressure drop experiments usually involve direct discharge of gas to atmosphere, a downstream pressure tap is rarely used, the downstream pressure being assumed as atmospheric. The location of the upstream pressure tap is more of a problem, the practice varying from placing it below the gas inlet aperture in the entrance pipe [228, 104, 164] to setting it slightly above the plane of the aperture [228, 134, 124, 129, 113]. When the column is run without solids at spouting gas rates, the former procedure produces a positive pressure drop, while the latter usually gives rise to a negative value, the suction being produced by a Venturi effect immediately above the entrance [13, 228]. Some investigators [124, 129, 228, 104] have therefore adopted the practice of algebraically subtracting the solids-free pressure drop from the measured bed pressure drop at the same mass flow rate of gas.

The same subtraction procedure applied to gas-fluidized bed pressure drop measurements in order to correct for the pressure drop across the grid has been criticized by Sutherland [222] on the grounds that at a given mass flow rate, the grid pressure drop is different in the presence of a bed than in the absence of a bed, due to elevation of the grid pressure level by the bed. Sutherland's derivation is based on the assumption that gas flow through the grid obeys Darcy's law for viscous flow through a porous medium, but in the case of a spouted bed this assumption would normally not apply to the high velocity jet inlet, even if a screen support were located in the aperture. Nevertheless, it is possible to generalize Sutherland's derivation to include nonviscous flow, as follows:

For isothermal flow of a gas through any confinement, the frictional pressure drop $P_1 - P_2$ can be expressed as the sum of a viscous term and an inertial term:

$$P_1 - P_2 = c_v \mu l u + c_i \rho_f u^2 l^2 / 2 \tag{2.3}$$

For a fixed confinement and fluid (assuming negligible variation of viscosity with pressure), Eq. (2.3) simplifies to

$$P_1 - P_2 = (c_v' G / \rho_f) + (c_i' G^2 / \rho_f) \tag{2.4}$$

G being the mass velocity of the fluid. Assuming the fluid is an ideal gas to the extent that ρ_f is proportional to P, it is permissible to write in good approximation

$$P_1 - P_2 = (c_v'' G + c_i'' G^2)/\tfrac{1}{2}(P_1 + P_2) \tag{2.5}$$

or

$$P_1{}^2 - P_2{}^2 = 2(c_v'' G + c_i'' G^2) \tag{2.5a}$$

Thus for a given mass flow rate, it is $P_1{}^2 - P_2{}^2$ and not $P_1 - P_2$ which will remain constant irrespective of the absolute pressure level. Hence the procedure of subtracting a measured positive pressure drop in the absence of solids from the measured value in the presence of solids, at the same mass flow rate of gas, leads to a negative error in the result. As an example, consider a spouting vessel open to the atmosphere (barometer reads 1040 cm of water) with a pressure tap, connected to a manometer, located at a point below the gas inlet orifice. If, for a given mass flow rate of gas, the manometer reads 40 cm of water with the vessel empty and 160 cm of water with a solids bed in the vessel, then the pressure drop due to the bed alone should be taken as $[(1200^2 - 1040^2) - (1080^2 - 1040^2) + 1040^2]^{1/2} - 1040 = 124$ cm of water and not as $160 - 40 = 120$ cm. Conversely, algebraically subtracting a measured negative pressure drop without solids from the positive value with solids will lead to a positive error, assuming that the Venturi effect at the vessel inlet also follows Eq. (2.3). For deep spouted beds in which the upstream tap is located in the entrance pipe significantly below the bed inlet aperture, the magnitude of such errors can be of the order of 10%, but normally the error is much smaller.

Some investigators who located their upstream pressure tap very close to the inlet aperture apparently made no correction whatsoever for the small pressure difference without solids [134, 113, 164]. On increasing the flow rate beyond the onset of spouting, Nelson and Gay [164], unlike almost all other workers, found the pressure drop to increase measurably. Since their upstream pressure tap was situated just below the inlet aperture, and since they did not apply any correction to account for pressure drop in the absence of solids, a plausible explanation for the observed pressure drop rise during spouting is that it was caused by the column entrance rather than by the bed. Other experimenters [71, 72, 79, 158–160, 168, 256, 15, 173] have reported neither the location of their pressure taps nor any correction procedure. Moderate discrepancies between the reported results of various investigations are therefore to be expected.

B. Peak Pressure Drop

The high peak in pressure drop which occurs just before spouting sets in is not a unique feature of a spouted bed but is associated generally with entry of a high velocity gas jet into a bed of solids. The occurrence of a similar peak has been reported for a channeling fluidized bed [116, p. 89], as well as for normal fluidization in both conical and conical-cylindrical vessels [71]. In these situations, as in spouting, the gas jet must first penetrate the solids in the lower region of the bed before causing movement of solids in the upper part. Even in a cylindrical fluidized bed where the gas enters through a uniform distributor, the same phenomenon occurs, but the excess pressure drop attained prior to fluidization is only slight, since each small gas jet which enters the solids through numerous orifices in the distributor can penetrate only a few particle layers before losing its identity by breaking into bubbles [263, p. 282]. The occurrence of a peak in the curve of pressure drop versus flow rate prior to the onset of both spouting and fluidization can, therefore, be attributed to the energy required by the gas jet to rupture the packed bed structure and to form an internal spout in the lower part of the bed. Whether this internal spout subsequently develops into a through-spout or gives rise to fluidization will depend on whether or not the critical

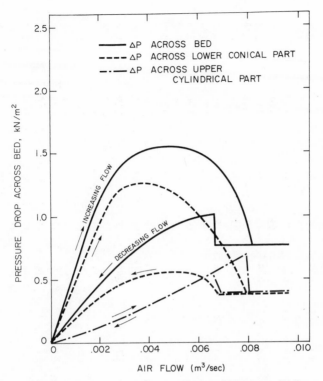

Fig. 2.5. Pressure drop–flow rate curves of Manurung [134].

conditions, such as particle size, orifice diameter, bed depth, etc., required for the spouting action are satisfied.

This explanation for the existence of a peak pressure drop is supported by the experimental results of Manurung [134], who measured pressure drops separately across the upper cylindrical part ($D_c \sim 15$ cm) and the lower conical part ($\theta = 60°$) of a rape seed bed contained in a conical-cylindrical column, as a function of both increasing and decreasing air flow (Fig. 2.5). It is seen that the pressure drop curves for the upper part of the bed, up to the point at which the spout starts to break through, correspond to those which are typically obtained in a packed bed [116, p. 50]. That they are practically coincident for both increasing and decreasing air flow below incipient spouting indicates that Manurung must have started his runs with the static bed in a loose packed condition, as was probably also done by most other investigators in the normal execution of their experiments. The curves for the lower part of the bed, on the other hand, are consistently far apart and show peaks well before incipient spouting.

TABLE 2.1 Peak Pressure Drop Data

Source	D_c (cm)	D_i (cm)	θ (deg)	Solids	d_p (mm)	γ (deg)	ρ_b (Mg/m³)	H (cm)	$-\Delta P_M$ observed (kN/m²)	Bed wt/area = $g\rho_b H$ [Eq. (2.6)] (kN/m²)	$-\Delta P_M$ by Eq. (2.7) (kN/m²)
Manurung [134]	15.2	1.27	60	Millet	1.3	54.5	0.77	34	2.72	2.57	2.76
								78	6.42	5.75	6.61
				Polystyrene	1.8	59.0	0.66	39	2.61	2.51	2.51
								64	4.37	4.10	4.29
				Coal	1.8	67.5	0.74	43	2.92	3.10	2.78
								59	4.45	4.43	4.16
								93	6.99	6.56	6.37
Lefroy and Davidson [114]	30.5	2.54	180	Kale seed	1.7	52[a]	0.58	15	0.93	0.85	0.73
								30	1.42	1.70	1.80
								60	3.73	3.41	3.93
								120	7.55	6.82	8.20
								180	11.77	10.24	12.50
Thorley et al. [228]	61.0	10.16	45	Wheat	4.8	55[b]	0.90	61	9.72	5.36	7.09
								180	28.92	15.75	23.65
Malek and Lu[c] [129]	15.2	0.95	60	Wheat	3.7	55[b]	0.83	28	2.15	2.25	1.44
								71	6.06	5.76	4.88
		2.54						35	2.71	2.86	1.83
								52	4.23	4.22	5.71
		3.81						28	2.20	2.26	3.48
								42	3.37	3.40	5.73
Nelson and Gay [164]	45.7	13.2	90	Peanuts	10.7	57[d]	0.16	22	5.74	3.52	1.48
								32	7.38	5.08	4.72
								42	10.38	6.70	8.06
								52	12.28	8.31	11.41
								62	16.07	9.92	14.74

[a] Assumed same as for Manurung's rape seed.   [b] From Lama [111].   [c] Pressure tap 2.54 cm above orifice.   [d] Measured by the present authors.

Despite the coincidence that both parts of the column show the same pressure drop in the stable spouting condition, the ascending curve for the lower part displays a much higher peak than the other three partial column curves. Apparently, then, most of the energy consumed by the gas jet in rupturing the mass of particulate solids is expended in the lower part of the bed on increasing the gas flow, no rupture energy being required for decreasing flow.

An early attempt by Madonna and Lama [124] to correlate the peak pressure drop $-\Delta P_M$ by a viscous flow packed bed equation of the Leva [116, p. 37] variety, allowing for the "rupture" pressure drop in the numerical value of the constant, was not too successful, since the "constant" itself was found to vary widely with column size as well as with particle properties. Moreover, their equation did not allow for the effect of gas inlet diameter. Subsequently, Malek and Lu [129], on the basis of data for several solid materials in columns of 10–30 cm diameter with bed depth to column diameter ratios greater than unity, arrived at the simple relationship that the maximum pressure drop approximately equals the weight of the bed per unit cross-sectional area. This relationship, first proposed by Becker [15] and most recently by Pallai and Németh [173], is equivalent to the statement

$$-\Delta P_M = H(\rho_s - \rho_f)(1 - \epsilon)g = H\rho_b g \qquad (2.6)$$

as for particulate fluidization, the buoyancy effect of the fluid being negligible in the case of moderate pressure gas flow. While Eq. (2.6) certainly gives a reasonable first approximation of the peak pressure drop, a check against some of the data reported by other workers (Table 2.1) shows that its prediction for $-\Delta P_M$ is usually on the low side, especially for larger columns. The generally lower results of Malek and Lu can perhaps be accounted for by the fact that these workers, unlike the others cited in Table 2.1, located their upstream pressure tap as high as 25 mm above the inlet orifice.

Another equation for predicting peak pressure drop has been developed by Manurung [134], who considered $-\Delta P_M$ to be composed of a rupture pressure drop and a frictional pressure drop. The latter is the total pressure drop for decreasing gas flow while the former is given by the difference between the total pressure drops for increasing and decreasing gas flow (see Fig. 2.5), both evaluated at the peak condition. Manurung confined his experimental work to two columns, each close to 15 cm in diameter with 60° cones, but studied a wide variety of solid materials consisting of both close fractions and mixed sizes (crushed coal, plastic particles, rape seed, millet, $d_p = 1$–4 mm, $\rho_s = 0.92$–1.43 Mg/m^3). From the experimental results he formulated empirical relationships for the two pressure drop components separately and then combined them to yield the following equation:

$$\frac{-\Delta P_M}{H \rho_b g} = \left[\frac{6.8}{\tan \gamma}\left(\frac{D_i}{D_c}\right) + 0.80\right] - 34.4 \frac{d_p}{H} \qquad (2.7)$$

The coefficient of internal friction, $\tan \gamma$, introduced to allow for the varying surface characteristics of the widely different solid materials, was measured according to the method of Zenz and Othmer [263, p. 75], and its value found to vary from 1.25 for rape seed to 3.2 for coal. For the mixed size particles, d_p was taken as the reciprocal mean diameter, $(\Sigma \ x_i/d_{pi})^{-1}$, determined from screen analysis data.

Table 2.1 shows that, excluding the aforementioned data of Malek and Lu, Eq. (2.7) gives better agreement in the main with the results of other workers than does Eq. (2.6), at least for H/D_c in excess of unity. Cone angle, which was not varied by Manurung in his experiments, does not appear in Eq. (2.7). Unless the cone angle is very small, it would not be expected to have any pronounced effect on jet penetration immediately above the inlet aperture. This view is supported by the reasonable agreement in Table 2.1 between $-\Delta P_M$ as predicted by Eq. (2.7) and the data of Lefroy and Davidson [114] for kale seed spouted with air in a flat-based column. On the other hand, any irreproducibility of the data due to possible bed-history dependence of $-\Delta P_M$ is not reflected in Eq. (2.7), so that correlation in the best of circumstances is unlikely to be perfect.

Nelson and Gay [164] have presented for the specific case of spouting whole peanuts a two-term empirical equation for peak pressure drop containing the same variables as Eq. (2.7) except for γ, which remains constant for a given solid material. By combining this equation with another empirical equation relating pressure drop and velocity prior to and including the peak condition, they were able to eliminate $-\Delta P_M$ and arrive at an explicit relationship for U_M, also specific to whole peanuts. Equations for U_M which attempt greater generality have been presented by Pallai and Németh [173]: For $H/H_m \leq 0.7$,

$$U_M/U_{mf} = 1 - (D_i^2/D_c^2)(0.7 - (H/H_m)) \qquad (2.8)$$

and for $H/H_m \geq 0.7$,

$$U_M/U_{mf} = 1 \qquad (2.9)$$

These equations require a knowledge of U_{mf}, the minimum fluidization velocity, and H_m, the maximum spoutable bed depth. U_{mf} can be estimated by substituting the pressure drop as given by Eq. (2.6) into a reliable packed bed relationship such as the Ergun equation [61] for the condition of $\epsilon = \epsilon_{mf} = \epsilon_0$ [54; 117, p. 21], while the prediction of H_m is discussed in

Chapter 6. Use of Eq. (2.8) or (2.9) thus allows estimation of the velocity at which the peak pressure drop occurs, without prior knowledge of the peak pressure drop itself.

The peak pressure drop in *conical* vessels has received the attention of several Soviet studies [71, 72, 79, 159, 168, 256]. Gelperin *et al.* [71, 72] obtained experimental values of $-\Delta P_M$ which in some cases were two to three times that necessary to support the bed weight. Their study was primarily concerned with fluidization of relatively fine particles, but their findings are considered to be relevant to spouting since the magnitude of the peak pressure drop should not depend on whether the bed will subsequently fluidize or spout. Their final empirical correlation was

$$-\Delta P_M/H\rho_b g = 1 + 0.062(D_b/D_i)^{2.54}((D_b/D_i) - 1)\cdot(\tan\theta/2)^{-0.18} \quad (2.10)$$

← conical

where θ is the included cone angle and D_b is the diameter of the upper surface of the bed. The range of variables covered was $\theta = 10°-60°$ and $H = 10-25$ cm, the inlet diameter having been restricted to 5 cm and the solid material to fine quartz of 0.16–0.28 mm size. The term D_b is related geometrically to the other conical bed dimensions by the equation

$$D_b = D_i + 2H \tan\theta/2 \quad (2.11)$$

Goltsiker *et al.* [79] contacted larger particles (3.2 mm diameter) with gas in conical vessels and obtained distinct spouting action beyond the point of peak pressure drop. The observed values of $-\Delta P_M$ were, however, lower than those predicted by Eq. (2.10). From a theoretical derivation for pressure drop across a packed bed, these workers showed that the absence of particle diameter in Eq. (2.10), though justified for laminar flow, is no longer valid for larger particles of the order of a few millimeters. This theoretical approach has recently been pursued further with some success by Wan-Fyong *et al.* [256].

Mukhlenov and Gorshtein [159], who also worked with conical vessels, argued that the ratio between the peak pressure drop and the pressure drop at steady spouting should bear a relationship to the geometry of the system, and to the properties of the gas and the solids. Starting with dimensional analysis, they arrived at the following empirical correlation of their experimental data:

$$\Delta P_M/\Delta P_s = 1 + 6.65(H/D_i)^{1.2}(\tan\theta/2)^{0.5}(\text{Ar})^{0.2} \quad (2.12)$$

where $-\Delta P_s$, the spouting pressure drop, is given by Eq. (2.29) and Ar, the Archimedes number, depends on the properties of the gas and the solids. The range of data supporting the above equation (within $\pm 10\%$) is as follows: $\theta = 12°–60°$, $D_i = 1.03–1.29$ cm, $H = 3–15$ cm, $d_p = 0.5–2.5$ mm, and $\rho_s = 0.98–2.36$ Mg/m³.

As a rough approximation for conical vessels generally, Nikolaev and Golubev [168] have stated that the ratio $\Delta P_M/\Delta P_s$ normally falls between 1.5 and 2.0.

C. Spouting Pressure Drop

In the spouted state the pressure drop across the bed arises out of two parallel resistances, namely, that of the spout in which dilute phase transport is occurring, and that of the annulus, which is a downward moving packed bed with counterflow of gas. Since the gas entering the base flares out radially from the axial zone into the annulus as it travels upward, the vertical pressure gradient increases from zero at the base to a maximum at the top of the bed. The total pressure drop across the bed can therefore be obtained by integrating the longitudinal pressure gradient profile over the height of the bed. Since the frictional pressure gradient for minimum fluidization,

$$(-dP/dz)_{mf} = (\rho_s - \rho_f)(1 - \epsilon_{mf})g \tag{2.13}$$

is approached only in the upper part of a deep bed, the total pressure drop across a spouted bed, $-\Delta P_s$, is always less than the pressure drop which would arise if the same solids were fluidized, given by Eq. (2.6). From theoretical considerations it will be shown that for the maximum spoutable bed depth, that is, for $H = H_m$, the spouting pressure drop bears a fixed ratio to the corresponding fluidization pressure drop $-\Delta P_F$ given by

$$-\Delta P_F = H_m(\rho_s - \rho_f)(1 - \epsilon_{mf})g \tag{2.14}$$

It was originally reasoned by Mathur and Gishler [137] that the pressure gradient in the annulus at any level depends on the upward gas velocity through the annulus at that level. Since the annulus is essentially a loose packed bed [58, 145], it follows that at any annular level

$$-dP/dz = KU_a^n \tag{2.15}$$

where n is a flow regime index which varies from 1 for viscous flow to 2 for fully turbulent flow [116, p. 49]. But at $z = H = H_m$, $U_a = U_{mf}$ and $(-dP/dz) = (-dP/dz)_{mf}$. Consequently

$$(-dP/dz)_{mf} = KU_{mf}^n \tag{2.16}$$

Assuming no significant change in flow regime, and hence of K and n, over the depth of the annulus, combination of Eqs. (2.13), (2.15), and (2.16) leads to

$$\frac{-dP/dz}{(\rho_s - \rho_f)(1 - \epsilon_{mf})g} = \left(\frac{U_a}{U_{mf}}\right)^n \tag{2.17}$$

A relationship similar to Eq. (2.17) was first formulated by Becker [15]. Its integration requires a knowledge both of n in the annulus and of the reduced annular velocity U_a/U_{mf} as a function of the reduced bed level z/H_m. Mamuro and Hattori [131] assumed $n = 1$, which transforms Eq. (2.15) to Darcy's law, and on this basis were able to derive a longitudinal velocity profile in the annulus,

$$U_a/U_{mf} = 1 - (1 - (z/H_m))^3 \tag{3.18}$$

as detailed in the next chapter. Substitution of Eq. (3.18) into Eq. (2.17) with $n = 1$ leads to

$$-dP = (\rho_s - \rho_f)(1 - \epsilon_{mf})g[1 - (1 - (z/H_m))^3] \, dz \tag{2.18}$$

Integration of Eq. (2.18) from $z = 0$ to $z = H_m$ yields

$$-\Delta P_s = (\rho_s - \rho_f)(1 - \epsilon_{mf})g(\tfrac{3}{4}H_m) \tag{2.19}$$

a result first obtained by Mamuro and Hattori [131]. Comparison of Eqs. (2.14) and (2.19) indicates that the ratio between the maximum value of spouting pressure drop and the corresponding fluidization pressure drop is equal to 0.75.

For practical spouted beds, the particle Reynolds number in the annulus is generally of the order of 100, for which n is considerably in excess of unity though still below 2 [117]. If the limiting value of 2 is assigned to n, then combination of Eq. (2.17) with Eq. (3.18) leads to

$$-dP = (\rho_s - \rho_f)(1 - \epsilon_{mf})g[1 - (1 - (z/H_m))^3]^2 \, dz \tag{2.20}$$

Although Eq. (3.18) was derived on the assumption that $n = 1$, its use for $n > 1$ can be justified by its reasonable agreement with experimental data (see Chapter 3) for which n in the annulus is in excess of unity [15]. Integration of Eq. (2.20) yields

$$-\Delta P_s = (\rho_s - \rho_f)(1 - \epsilon_{mf})g((9/14)H_m) \tag{2.21}$$

In this case, then, the maximum ratio of the spouting to the fluidization pressure drop is 9/14 or 0.643.

This lower value of the ratio is almost identical to that derived by Lefroy and Davidson [114] by a more direct approach which requires no knowledge of or assumptions about velocities. These investigators noted that the

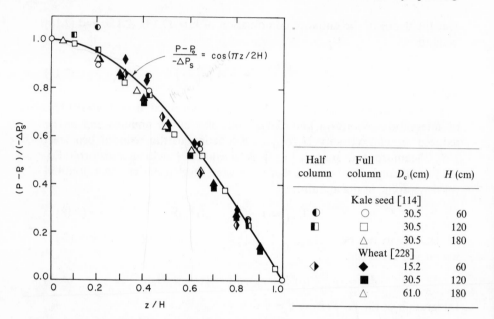

Fig. 2.6.   Longitudinal pressure profiles in spouted beds.

longitudinal pressure distributions obtained on spouting kale seed in their
flat-based column could be described by a quarter cosine curve (see Fig.
2.6), which can be generalized as

$$(P - P_0)/-\Delta P_s = \cos(\pi z/2H) \qquad (2.22)$$

where the static pressure at no flow, P_0, is essentially the bed exit pressure
unless the fluid is a liquid. Although they made their pressure measurements
just outside the spout, Lefroy and Davidson were able to demonstrate
theoretically that for most practical spouted beds, since H/D_c exceeds 2,
radial pressure gradients can be neglected. Equation (2.22) is therefore
equally applicable to pressure distributions measured at the column wall.
The generality of this representation is supported by Fig. 2.6, in which the
same equation is shown to fit wall pressure distributions for spouted beds
of wheat in columns of different diameters.

Differentiating Eq. (2.22) with respect to z, we obtain

$$-\frac{dP}{dz} = -\Delta P_s \frac{\pi}{2H} \sin \frac{\pi z}{2H} \qquad (2.23)$$

Putting $z = H = H_m$ and noting that for this condition the pressure gradient is sufficient to support the solids at the top of the annulus and is therefore given by Eq. (2.13), then it follows from Eq. (2.23) that

$$-\Delta P_s = (2H_m/\pi)(\rho_s - \rho_f)(1 - \epsilon_{mf})g \qquad (2.24)$$

In other words, the maximum ratio of spouting to fluidizing pressure drop is here derived as $2/\pi$ or 0.637.

These derived values of $\Delta P_s/\Delta P_F$ represent the upper limit which would be approached with increasing bed depth for a given system. Their validity is borne out by the experimental results plotted in Fig. 2.7 covering different solid materials and column geometries. The maximum ratios of $\Delta P_s/\Delta P_F$ attained are seen to be in remarkably good agreement with the predicted values of 0.64–0.75. Also in agreement with these values is the empirical equation of Pallai and Németh [173],

$$\frac{(\Delta P_s)_{max}}{\Delta P_F} = 0.8 - 0.01 \frac{D_{column}}{D_{inlet}} \qquad (2.25)$$

especially when it is realized that for practical spouted beds, D_c/D_i generally falls between 5 and 15.

For beds in which $H < H_m$, theoretical analysis is impeded by the difficulty of selecting a suitable upper boundary condition for integrating the differential pressure gradient. An attempt by Madonna and Lama [124]

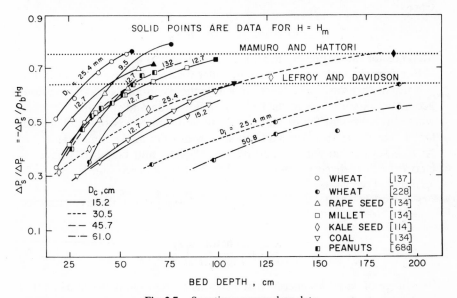

Fig. 2.7. Spouting pressure drop data.

to correlate spouting pressure drop with mass velocity using a Leva-type packed bed equation [116, p. 49], similar to their attempted ΔP_M correlation mentioned in the previous section, met with only limited success since it again failed to take into account the effect of column and inlet diameters. The simple suggestion of Pallai and Németh [173] that $-\Delta P_s$ is linear with H also overlooks these two variables, which do undoubtedly play a role (see Fig. 2.7). The form of the experimental curves in Fig. 2.7 is correctly described by an empirical equation due to Manurung [134],

$$\frac{-\Delta P_s}{H \rho_b g} = \frac{1}{1 + (0.81(\tan \gamma)^{1.5}/\bar{\psi}^2)(D_c d_p/D_i^2)^{0.78}(D_c/H)} \qquad (2.26)$$

which was based on the same range of data as Eq. (2.7). However, actual values predicted by Eq. (2.26) for large diameter beds of wheat (30.5 and 61 cm) and kale seed (30.5 cm) turn out to be higher than observed results by up to 30%. Prediction is improved if the use of Eq. (2.26) is restricted to calculating the spouting pressure drop for a given bed depth from its known value at the maximum spoutable bed depth. The relevant relationship is then

$$\frac{\Delta P_s}{(\Delta P_s)_{max}} = \frac{H}{H_m} \frac{1 + K'(D_c/H_m)}{1 + K'(D_c/H)} \qquad (2.27)$$

where

$$K' = (0.81(\tan \gamma)^{1.5}/\bar{\psi}^2)(D_c d_p/D_i^2)^{0.78} \qquad (2.28)$$

Methods for estimating H_m are discussed in Chapter 6.

For flow through packed beds an Euler number Eu or a friction factor, f is usually expressed as a function of a Reynolds number Re and of various geometrical parameters of the bed. For fluidization and spouting, however, where gravitational forces play a role in addition to viscous and inertial forces, one would also expect a Froude number Fr or an Archimedes number Ar to enter into any general correlation. Nelson and Gay [164] have included Fr in their correlation for $-\Delta P_s$, but their expression is specific to spouting of whole peanuts.

For *conical* vessels, Mukhlenov and Gorshtein [158, 160] have proposed that

$$\frac{-\Delta P_s}{H \rho_b g} = \frac{7.68(\tan \theta/2)^{0.2}}{Re_i^{0.2}(H/D_i)^{0.33}} \qquad (2.29)$$

The range of variables encompassed by this correlation is the same as that for Eq. (2.12). The Reynolds number Re_i is based on particle diameter and gas velocity through the inlet orifice.

2.3 MINIMUM SPOUTING VELOCITY

The minimum fluid velocity at which a bed will remain in the spouted state (U_{ms}) depends on solid and fluid properties on the one hand and bed geometry on the other. In a cylindrical column, U_{ms} for a given material, unlike U_{mf}, increases with increasing bed depth and with decreasing column diameter, as illustrated in Fig. 2.8. The size of the fluid inlet also has an effect on U_{ms}, though this is relatively small. Comparison with the minimum fluidization velocity is therefore difficult, except at the maximum spoutable depth H_m where U_{ms} reaches its maximum value.

A. Maximum Value

The value of U_{ms} at H_m, termed U_m or *the maximum of the minimum spouting velocity* [15] for a given material, might be expected to coincide with its minimum fluidization velocity, since beyond H_m a spouted bed transforms into a fluidized bed (Fig. 1.5). Experimental data, however, show that the connection between U_m and U_{mf} is more tenuous, with U_m often exceeding U_{mf}.

According to the data in Fig. 1.5, U_m does approximately equal U_{mf} in the case of sand, while it is higher than U_{mf} by 33% for wheat and 45% for

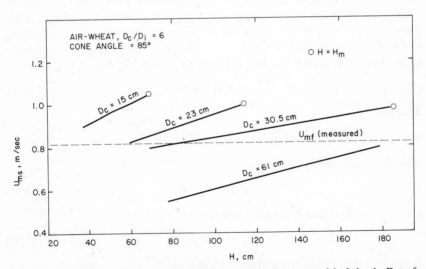

Fig. 2.8. Minimum spouting velocity: Effect of column diameter and bed depth. Data for $D_c = 15, 23$, and 30.5 cm: Mathur and Gishler [137]; data for $D_c = 61$ cm: Thorley *et al.* [228].

TABLE 2.2   Effect of Column and Orifice Size on U_m

Wheat ($d_p = 3.2$ mm), $U_{mf} = 0.82$ m/sec, $D_c/D_i = 6$, data from Fig. 2.8			Polythene chips ($d_p = 3.5$ mm), $U_{mf} = 0.97$ m/sec, $A_c = 61$ cm^2 (square column), data of Lefroy and Davidson [114]		
D_c (cm)	U_m (m/sec)	U_m/U_{mf}	D_i (cm)	U_m (m/sec)	U_m/U_{mf}
15.2 (6 in.)	1.05	1.28	5.1	0.95	0.97
22.9 (9 in.)	1.00	1.22	7.6	1.04	1.07
30.5 (12 in.)	0.95	1.16	10.2	1.12	1.15
			12.7	1.17	1.20

semicoke. Values of U_m exceeding U_{mf} by 10–33% have been reported by Becker [15] for a variety of uniform size materials, though his percentage figures are only approximate since he used calculated rather than observed values of U_{mf} for the comparison. Pallai and Németh [173], who were guided by their experiments in a 6 cm diameter column, have suggested that as a general rule $U_m \sim 1.5 U_{mf}$. The abovementioned variations in the ratio U_m/U_{mf} may be attributed either to differences in the properties of the solid materials or to the effect on U_m of the spouting vessel geometry. The dependence on solid properties is clearly brought out by the results mentioned above for sand and wheat, which were obtained in the same apparatus, while the data in Table 2.2 show that both column and orifice size influence U_m, and therefore U_m/U_{mf}. It is seen from Table 2.2 that for a fixed D_c/D_i ratio, U_m decreases with increasing column diameter, while for a fixed value of D_c (or column cross section) it increases with increasing orifice diameter. Both these trends are predictable from established empirical equations for U_{ms} and H_m. For a fixed value of D_c/D_i, Eq. (2.38) gives

$$U_{ms} \propto (1/D_c)H^{1/2} \tag{2.30}$$

Therefore

$$(U_{ms})_{Hm} = U_m \propto (1/D_c)H_m^{1/2} \tag{2.31}$$

while from Eq. (6.5),

$$H_m \propto D_c^{1.75} \tag{2.32}$$

Eliminating H_m from Eqs. (2.31) and (2.32), we get

$$U_m \propto 1/D_c^{0.125} \tag{2.33}$$

It can be similarly shown that for a fixed D_c,

$$U_m \propto D_i^{0.133} \tag{2.34}$$

The data in Table 2.2 are in surprisingly good agreement with the above relationships.

Thus, the precise relationship between U_m for a spouted bed and U_{mf} depends on both the properties of the solid material and the geometry of the spouting vessel, but the ratio U_m/U_{mf} does not vary widely. Values of this ratio for a large variety of materials in columns of 15 cm to 61 cm diameter have all been found to fall within the range 1.0–1.5.

A correlation for estimating U_m which neglects its dependence on column geometry has been proposed by Becker [15]:

$$C_D\psi = (2600/\text{Re}_m) + 22 \tag{2.35}$$

where

$$C_D = \frac{4d_v g(\rho_s - \rho_f)}{3\rho_f U_m{}^2}$$

and

$$\text{Re}_m = d_v U_m \rho_f/\mu$$

d_v being particle size expressed as diameter of an equivolume sphere. Equation (2.35) is based on extensive data for spouting of various uniform size materials with air in 15–61 cm diameter columns. Experimental results for U_m, when plotted as C_D versus Re_m, gave parallel curves for the different materials. A shape factor ψ was therefore introduced to modify C_D, the values assigned to it (1.0 for spheres; 0.62 and 0.76, respectively, for two different types of wheat; 0.35 for flax seed; etc.) being such that the data for particles of different shapes were brought together.

Although Becker's empirical equation [Eq. (2.35)] does correlate specific data on U_m, Manurung [134] has pointed out that its form is nevertheless similar to the relationship for U_{mf} based on Ergun's packed bed equation. Putting $-\Delta P/H$ in the Ergun equation equal to $(\rho_s - \rho_f)(1 - \epsilon_0)g$ yields the familiar fluidization velocity equation [110, p. 73]

$$1.75\rho_f d_p U_{mf}{}^2 + 150\mu(1 - \epsilon_0)U_{mf} - d_p{}^2\epsilon_0{}^3(\rho_s - \rho_f)g = 0 \tag{2.36}$$

which can be rearranged into the form

$$C_D = \frac{200(1 - \epsilon_0)}{\text{Re}_{mf}\epsilon_0{}^3} + \frac{7.0}{3\epsilon_0{}^3} \tag{2.37}$$

Equation (2.37), with a constant value of voidage, becomes similar to Eq. (2.35). Manurung's own results for U_m of coal, rape seed, millet, and polyethylene cubes obtained in a 15 cm diameter column showed better agreement with U_{mf} calculated from Eq. (2.36) than with U_m from Eq. (2.35).

TABLE 2.3 Minimum Spouting Velocity Correlations[a]

Investigators	Correlation	Range of supporting data	
		Bed geometry	Solids used
	Cylindrical vessels (with short conical base)		
Manurung [134]	$U_{ms} = 7.73(\tan\gamma)^{0.72}(d_p/D_c)^{0.62}(D_i/D_c)^{0.155\,\tan\gamma}(HU_m g)^{1/3}$ (2.45) $\tan\gamma$ = coefficient of internal friction measured by the method of Zenz and Othmer [263], U_m from Eq. (2.36). For mixed sizes, $d_p = 1/\sum(x_i/d_{pi})$. Introduction of a shape factor and a surface roughness factor in Eq. (2.36) was necessary to deal with irregular and rough coal particles. Equation found to be good for coal mixtures, but over a wider range of conditions, including previous data for wheat in columns of different size, it showed no clear improvement over Mathur–Gishler and Becker equations.	$D_c = 15$ cm $D_i = 9$–15.2 mm $H = 20$–100 cm $\theta = 60°$	Coal of six different sizes, both close fractions and mixtures, polystyrene, rape seed, and millet. $d_p = 1$–4 mm $\rho_s = 1.09$–1.43 Mg/m³
Smith and Reddy [219]	$U_{ms} = d_p\left(\dfrac{g(\rho_s - \rho_f)}{\rho_f D_c}\right)^{1/2}\left[\,0.64 + 26.8\left(\dfrac{D_i}{D_c}\right)^2\,\right]\left(\dfrac{H}{D_c}\right)^{0.5 - 1.76 D_i/D}$ (2.46) $d_p = \sum[x_i/d_{pi}]/\sum[x_i/d_{pi}^2]$ Good for mixed sizes but not for closely sized materials even in 15 cm diameter column. Found inapplicable when tested against data for 61 cm diameter wheat beds.	$D_c = 15$ cm $D_i = 9$–19 mm $H = 33$–58 cm $\theta = 60°$	Alundum, sand, crystolon, and polystyrene, all with particle size spread spread.
Charlton et al. [36]	$U_{ms} = kV^{0.6}\rho_s^{0.6}d_p D_i^{0.2}$ (2.47) Units: cm-gm-sec V is the bed volume. Values of constant k found to be 10 and 12 for two different configurations of the conical base. Values of exponents on V and D_i depended on the other parameters; figures shown are averages for the range of conditions studied. Shallowest beds used did not extend beyond the conical section.	$D_c = 7.5$ cm $D_i = 1.27$–9.5 mm $H = 2.5$–20 cm $\theta = 30°, 60°$	Glass, steel, copper, and lead spheres, uniform size. $d_p = 0.5$–6.4 mm $\rho_s = 2.6$–11.0 Mg/m³ Fluids: air, carbon dioxide, and helium.

Author	Equation	Conditions	Particles
Abdelrazek [1]	(2.48) $$U_{ms} = \frac{1}{1.74}\left[\left(\frac{d_p}{D_c}\right)\left(\frac{D_i}{D_c}\right)^{1/3}\left(\frac{2gH(\rho_s - \rho_f)}{\rho_f}\right)^{1/2} - 0.25\right]$$ Units: ft-lb-sec. Modified version of Eq. (2.38) applicable only to small beds of fine particles.	$D_c = 5$–10 cm $D_c/D_i = 12$ $H/D_c = 1$–3 $\theta = 60°$	Glass and steel spheres, uniform size. $d_p = 0.5$–0.8 mm $\rho_s = 2.46$ and 7.07 Mg/m^3
Pallai and Németh [173]	(2.49) $$U_{ms} = U_{mf}\left[\frac{H/H_m}{1.5} + 1.0\right]$$ Implies that U_{ms} always exceeds U_{mf}, and that $U_m = 1.67$; therefore, of limited applicability (see Fig. 2.9 and Section 2.3A).	$D_c = 6$ cm $D_i = 6, 8, 10$ mm $\theta \sim 60°$	Glass spheres and activated charcoal. $d_p = 1.6$–2.5 mm
Gay et al. [68]	(2.50) $$U_{ms} = 1.26\times10^7\,\frac{d_p\rho_b g}{\rho_f}\left(\frac{H}{D_c}\right)^{0.19(D/D)1.38}\left(\frac{D_i}{D_c}\right)^{5.62}\left(\frac{\mu^2}{\rho_b\rho_f D_c^3\,g}\right)^{0.5}$$ d_p taken as average diameter of the kernel on the basis of free fall experiments.	$D_c = 38$–61 cm $D_i = 3.8$–13.2 cm $H = 18$–62 cm $\theta = 90°$	Whole peanuts, 11 mm diameter × 23 mm. $\rho_b = 0.32$ Mg/m^3

Conical vessels

Author	Equation	Conditions	Particles
Nikolaev and Golubev [168]	(2.51) $$(Re)_{ms} = 0.051(Ar)^{0.59}(D_i/D_c)^{0.10}(H/D_c)^{0.25}$$	Cone plus short cylinder, 12 cm diameter. $D_i = 2$–5 cm $H = 9$–15 cm $\theta =$ not given	Spherical particles of five different sizes. $d = 1.75$–5.6 mm
Gorshtein and Mukhlenov[b] [84]	(2.52) $$(Re_i)_{ms} = 0.174\,\frac{(Ar)^{0.50}}{(\tan\theta/2)^{1.25}}\left(\frac{D_b}{D_i}\right)^{0.85}$$ where $Re_i = d_p u_i \rho_f/\mu$.	Conical vessel with a short cylindrical upper part. $D_i = 1.0$–1.3 cm $H = 3$–15 cm $\theta = 12°$–$60°$	Quartz, sand, millet, aluminum silicate. $d_p = 0.5$–2.5 mm $\rho_s = 0.98$–2.36 Mg/m^3 $\rho_b = 0.70$–1.63 Mg/m^3

TABLE 2.3 (continued)

Investigators	Correlation	Range of supporting data	
		Bed geometry	Solids used
Tsvik et al. [230]	(2.53) $(Re_i)_{ms} = 0.4(Ar)^{0.52}(H/D_i)^{1.24}(\tan\theta/2)^{0.42}$	$D_i = 2.0-4.2$ cm $H = 10-50$cm $\theta = 20°-50°$	Fertilizer fractions. $d_p = 1.5-4.0$ mm $\rho_s = 1.65-1.70$ Mg/m^3 $\rho_b = 0.78-0.84$ Mg/m^3
Goltsiker [78]	(2.54) $(Re_i)_{ms} = 73(Ar)^{0.14}(\rho_s/\rho_f)^{0.47}(H/D_i)^{0.9}$ Equation proved valid for antimony sand with size spread between 0.6 and 5.0 mm, using the reciprocal mean diameter as d_p.	$D_i = 4.1-12.3$ cm $H = 5-31$ cm $\theta = 26°-60°$	Fertilizer and silica gel, closely sized. $d_p = 1.0-3.0$ mm
Wan-Fyong et al. [256]	(2.55) $(Re_i)_{ms} = k\,Re_t(H/D_i)^{0.82}(\tan\theta/2)^n$ where Re_t is Reynolds number at terminal velocity of particle. For θ between 16° and 70°, $k = 1.24$ and $n = 0.92$; for θ between 10° and 16°, $k = 0.465$ and $n = 0.49$.	Cone plus short cylinders, 11.2–20 cm diameter. $D_i = 2.6-7.6$ cm $H = 7-30$ cm $\theta = 10°-70°$	Millet, silica gel, four different plastics; close cut as well as with up to fourfold size spread. $d_p = 0.35-4$ mm. $\rho_s = 0.45-1.39$ Mg/m^3 $\rho_b = 0.20-0.79$ Mg/m^3

[a] Spouting fluid, air (except in investigation of Charlton et al. [36]).
[b] Two modified versions of this correlation have been subsequently published by the same workers [159, 160].

He suggested that the implied assumption of a constant packed bed voidage in the latter equation makes it less reliable than Eq. (2.36), since ϵ_0 does vary considerably with particle shape, size, and size distribution. This explanation is supported by the fact that measured values of ϵ_0 for the materials used by Manurung in testing Eq. (2.35) varied as widely as 0.37 to 0.53, as against 0.37 to 0.41 for Becker's solids. Manurung's results for crushed coal, which had rough and irregular particles, were not, however, correlated even by Eq. (2.36).

With the limitations of Eq. (2.35) brought out by Manurung's work, it becomes evident that this equation, though specifically developed for U_m, is not sufficiently sensitive to distinguish between U_m and U_{mf}. Indeed, the absence of a voidage term and the presence of an experimental shape factor make Eq. (2.35) even less reliable than Eq. (2.36) as a general predictive correlation for U_m.

In conclusion, it should be pointed out that a knowledge of U_m per se is not of much practical value, since the operating bed depth in spouting units of industrial size would usually be less than the maximum spoutable depth. The concern with U_m, and its relationship to U_{mf}, arises mainly from its possible usefulness as a reference parameter in developing correlations for U_{ms} and for H_m, since U_m is far more independent of column geometry than both U_{ms} and H_m.

B. Correlations

For bed depths of practical interest, e.g., with $H/D_c = 2–4$, the minimum spouting velocity for a particular material can be either higher or lower than its minimum fluidization velocity, depending on the scale of operation, as brought out by the data in Fig. 2.8. The important problem of developing a correlation for predicting U_{ms}, which has received wide attention [1, 15, 36, 50, 68, 75, 78, 84, 104, 130, 134, 137, 168, 173, 219, 228, 230, 256] is, therefore, inherently more complex than the corresponding problem in either fluidization or pneumatic transport, where the minimum velocities are independent of scale. Experimental data on U_{ms} are abundant, covering a wide range of materials in small columns of both cylindrical and conical shapes. The effect of column size up to 61 cm diameter has also been studied, but mostly with wheat beds. Because of the complexity of the system, the approach to correlating experimental data has been almost entirely empirical, with the result that over a dozen different spouting velocity equations have appeared in the literature. In the absence of any unifying theoretical analysis of the problem, this is perhaps not surprising. Only two of these equations for cylindrical vessels are considered sufficiently general to be of practical value and warrant further discussion; all the others for both cylindrical and conical vessels are listed in Table 2.3, together with the range of supporting data and brief comments in each case.

1. Mathur–Gishler Equation*

$$U_{ms} = (d_p/D_c)(D_i/D_c)^{1/3}(2gH(\rho_s - \rho_f)/\rho_f)^{1/2} \qquad (2.38)$$

The above equation, developed with the help of dimensional analysis, was derived from results for a number of closely sized materials spouted in 7.6 to 30.5 cm diameter columns using air as well as water as the spouting fluids. Over the years, it has proved to be valid for a much wider range of conditions (see Table 2.4), which include not only a larger variety of solid materials, closely sized as well as with size spread, but also column diameters up to 61 cm. Although no theoretical basis for Eq. (2.38) was claimed by its authors, Ghosh [75] subsequently arrived at a similar relationship by equating the momentum gained by the particles with that lost by the entering fluid. He assumed that a particle at the bottom must attain a velocity equal to $(2gH)^{1/2}$ in order to reach the top of the spout, and that the number of particles following each other in succession per unit time is proportional to v/d_p, since each particle requires a time interval of d_p/v to vacate its position. Considering that n particles can enter the spout through its periphery at any instant, the total number of particles accelerated per unit time becomes $n(v/d_p)$. Hence the momentum gained by the particles (assuming spherical shape) per unit time is

$$M = (nv/d_p)(\pi d_p{}^3/6)(\rho_s - \rho_f)(2gH)^{1/2} \qquad (2.39)$$

where $v = (2gH)^{1/2}$. The momentum (per unit time) lost by the fluid jet is proportional to its initial momentum (per unit time) and is therefore given by

$$M = K(\pi D_i{}^2/4)u_i\rho_f u_i \qquad (2.40)$$

where u_i is the jet velocity through the orifice at minimum spouting. Solving Eqs. (2.39) and (2.40) for u_i, and converting to U_{ms} by multiplying both sides by $(D_i/D_c)^2$, we have

$$U_{ms} = \left(\frac{2n}{3K}\right)^{1/2}\left(\frac{d_p}{D_c}\right)\left(\frac{D_i}{D_c}\right)\left(\frac{2gH(\rho_s - \rho_f)}{\rho_f}\right)^{1/2} \qquad (2.41)$$

Aside from the numerical constant, the only point of difference between Eq. (2.41) and Eq. (2.38) is the exponent on the group D_i/D_c, its value being $\frac{1}{3}$ in the empirical equation as against unity in the theoretical. This difference

*See Mathur and Gishler [137].

TABLE 2.4 Range of Conditions over Which Eq. (2.38) Has
Proved to Be Valid within about $\pm 10\%$

Column diameter (cm)	Bed solids		ρ_s (Mg/m^3)	U_{ms}, expt (m/sec)	Ref
	Material	Size			
Spouting fluid, air; $H/D_c = 1.3$–6.7, $D_c/D_i = 3.3$–24					
	Closely sized				
15	Wheat	3.2 mm × 6.4 mm	1.38	0.74–1.07	[137]
	Mustard seed	2.2 mm	1.20	0.43–0.77	
	Rape seed	1.8 mm	1.10	0.35–0.62	
	Peas	6.4 mm	1.39	1.26–1.62	
	Ottawa sand	−20 +30 mesh	2.32	0.17–0.23	
	Gravel	− 4 + 8 mesh	2.63	1.42	
		−14 +20 mesh		0.52	
	Millet	1.3 mm	1.29	0.28–0.48	[134]
	Nylon flakes	− 4 +10 mesh	1.10	0.59–1.10	
	Polystyrene	− 6 + 9 mesh	1.05	0.70–0.82	
		− 9 +16 mesh		0.28–0.43	
	Coal, rounded	− 6 +12 mesh	1.43	0.64–0.93	
		− 8 +14 mesh		0.49–0.82	
		−10 +35 mesh		0.21–0.46	
	Coal, irregular	− 8 +14 mesh	1.43	0.40–0.73	
		− 9 +24 mesh		0.24–0.49	
		−10 +28 mesh		0.18–0.37	
	Alundum	−35 +48 mesh	3.95	0.15–0.18	[219]
23	Wheat	3.2 mm × 6.4 mm	1.38	0.70–1.16	[137]
30.5	Wheat	3.2 mm × 6.4 mm	1.38	0.61–0.98	
61	Wheat	3.6 mm^a (smaller dimension)	1.38	0.51–0.76	[228]
	Mixed size				
15	Coal, rounded (13 mixtures)	− 6 +35 mesh	1.43	0.27–0.57	[134]
	Coal, irregular (4 mixtures)	− 6 +28 mesh	1.43	0.27–0.59	
	Polystyrene (3 mixtures)	− 6 +16 mesh	1.05	0.28–0.65	
	Alundum (11 mixtures)	− 9 +48 mesh	3.95	0.23–1.28	[219]
	Crystolon	−16 +32 mesh	3.20	0.28–0.47	
	Silica sand	−10 +28 mesh	2.64	0.34–0.56	
	Polystyrene (5 mixtures)	1.1 × 1.4 × 2.1 mm^3 plus 1.8 × 2.7 × 3.7 mm^3	1.05	0.36–0.87	

TABLE 2.4 (continued)

Column diameter (cm)	Bed solids		ρ_s (Mg/m^3)	U_{ms}, expt (m/sec)	Ref
	Material	Size			
	Spouting fluid, water; $H/D_c = 1–4.7$, $D_c/D_i = 6–48$				
7.6; 15	Quartz (crushed)	$-14 +20$ mesh	2.63	0.009–0.012	[137]
		$-20 +35$ mesh		0.003–0.006	
	Catalyst spheres	$-4 + 6$ mesh	3.51	0.039–0.051	
		$-6 + 8$ mesh		0.013–0.037	
	Alundum	$-10 +14$ mesh	3.88	0.013–0.035	
	Ottawa sand	$-20 +35$ mesh	2.32	0.003–0.011	
	Gravel	$-8 +14$ mesh	2.63	0.013–0.016	
		$-20 +35$ mesh		0.002–0.007	
	Brucite	$-8 +14$ mesh	2.51	0.011–0.020	
		$-14 +20$ mesh		0.005–0.023	
	Galena	$-20 +35$ mesh	7.44	0.008–0.016	

[a] Thorley *et al.* have not reported the sizes of their wheat particles. The value of 3.6 mm was back-calculated [220] from the U_{ms} (calc) results in Table 2 of [228].

is not surprising since a substantial portion of the fluid entering the orifice flares out into the annulus immediately above the orifice, so that the momentum gained by the solids is only weakly related to the initial momentum rather than being directly proportional to it. Besides, the term n is likely to be a function of D_i/D_c, as Ghosh himself pointed out.

The following comments on the use of Eq. (2.38) and its limitations are necessary:

(1) Correct choice of the characteristic particle dimension d_p is important. For closely sized materials consisting of near spherical particles, d_p should be taken as the arithmetic mean of screen apertures, while for beds consisting of mixed size particles, the equation requires the use of reciprocal mean diameter [220],

$$d_p = 1/\Sigma(x_i/d_{pi})$$

Here x_i represents the weight fraction of particles of size d_{pi} and the value of d_{pi} for each size cut can again be taken as the arithmetic average of screen apertures. If the particle shape deviates widely from the spherical, however, a value of d_p based on average screen aperture may become unsuitable. Thus experimental data for wheat are best correlated when d_p is taken as the smaller dimension of the particle, this choice being justified by the observation that wheat particles align themselves vertically in the spout [137]. For other shapes, no general criterion for selecting the characteristic dimension can be given, although the use of an equivolume sphere diameter has proved to be effective for certain regular shaped nonspherical particles [134]. In practice, the most reliable and convenient method would be to determine an effective value of d_p empirically in a laboratory spouting unit. An additional advantage of this procedure is that the surface characteristics of the particles, which can also influence U_{ms} [134] but are not explicitly accounted for in Eq. (2.38), and are in any case difficult to define quantitatively, will be reflected in the effective value of d_p.

(2) The fluid density dependence in Eq. (2.38) is based on results for air and water, but the effect of fluid properties if the fluid is a gas other than air has not been fully established. The absence of fluid viscosity from the equation has been questioned in particular [36]. Some data obtained in a 15 cm diameter column using beds of wheat, barley, peas, and coffee beans spouted with superheated steam [180], which entered the bed at about 200°C and left at 120–150°C, showed good agreement with Eq. (2.38). Fluid density (~ 0.03 Mg/m³) was evaluated at the arithmetic average of the inlet and outlet temperatures. Thus, no viscosity effect was revealed in these tests, the viscosity of superheated steam at the temperatures used being about 50% higher than that of air at room temperature.

(3) The included angle of the conical base, which varied between 30° and 85° for the data in Table 2.4, did not significantly affect the spouting velocity for columns up to 30.5 cm diameter. In a 61 cm column, however, Thorley *et al.* [228] found the spouting velocity for wheat to be about 10% higher with an 85° cone than with a 45° cone. The large column results were better correlated if the exponent to the ratio D_i/D_c in Eq. (2.38) was reduced to 0.23 for 45° and 60° cone angles and to 0.13 for 85° [227].

2. Becker Equation*

$$U_{ms} = U_m[1 + s \ \ln(H/H_m)] \tag{2.42}$$

U_m is obtainable from Eq. (2.35) with a knowledge of the particle and fluid properties, while the column geometry variables are incorporated in coefficient s, which is given by the following empirical equation:

$$s = 0.0071(D_i/D_c) \, \mathrm{Re}_m^{0.295} \psi^{2/3} \tag{2.43}$$

The value of H_m for substitution in Eq. (2.42) needs to be calculated by yet another empirical equation:

$$[H_m/d_v][d_v/D_c][12.2D_i/D_c]^{1.6 \ \exp(-0.0072\mathrm{Re}_m)}$$
$$\times \ [(2600/\mathrm{Re}_m) + 22]\psi^{2/3} \, \mathrm{Re}_m^{1/3} = 42 \tag{2.44}$$

Calculation by this method, according to Becker, is valid for Re_m of 10–100, H/D_c greater than 1, and D_i/D_c less than 0.1.

*See Becker [15].

The above set of equations was developed by Becker on the basis of not only his own data but also the previous data of Mathur and Gishler. It should be noted, however, that the equations are entirely empirical, and that since Becker's data covered more or less the same range of variables as those of the previous workers (in fact, a narrower range, inasmuch as only air was used as the spouting fluid), the generality of his calculation method is no better than that of Eq. (2.38). As for accuracy, Becker himself gives no indication as to how closely the predicted values of U_{ms} agreed with experimental results, nor did he make any comparison of his equation with the previous equation of Mathur and Gishler. Such a comparison was subsequently made by Manurung [134] using experimental data for a large variety of materials but only for a 15 cm diameter column. He noted that the effect of increasing bed depth was somewhat overestimated by Eq. (2.38) and underestimated by Becker's equation, but the comparison did not clearly establish which of these two equations is the more reliable. Thus, as far as small diameter beds are concerned, the considerable complexity of Becker's correlation does not seem to have yielded any compensating benefits. Calculations done by the present authors show that the dependence of U_{ms} on column diameter in Eq. (2.38) is significantly stronger than in Eq. (2.42), and there is some indication that for $D_c > 61$ cm the spouting velocity might be better predicted by the latter equation, as discussed in the following section.

C. Prediction for Large Beds

Although the experimental results discussed in the previous section include data for columns up to 61 cm diameter, it should be recalled that most of the results for beds larger than 15 cm diameter are only for wheat. Two questions arise:

(1) Would the preferred spouting velocity equations [Eqs. (2.38) and (2.42)] apply to other solid materials in columns up to 61 cm diameter?

(2) With what confidence can these equations be extrapolated for predicting U_{ms} in larger columns?

In an attempt to answer the first question, we have checked only the simpler of the two equations [Eq. (2.38)] against data from the industrial grain drier listed in Table 11.2. Although the available data are not sufficiently precise for close verification, the operating air velocities for 61 cm diameter beds of near spherical, uniform size particles, namely, lentils ($d_p = 4.4$ mm) and peas ($d_p = 6.7$ and 7.5 mm), were found to be consistent

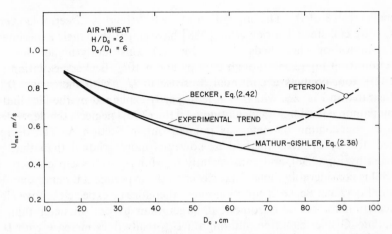

Fig. 2.9. Effect of scale-up on minimum spouting velocity: Prediction versus observation. Peterson's result for $H = 119$ cm extrapolated to $H = 183$ cm (i.e., $H/D_c = 2$) by multiplying his observed value by $(183/119)^{\frac{1}{2}}$.

with values predicted from Eq. (2.38), taking air density at the arithmetic average of inlet and exit temperatures.

The second question is more difficult to answer since, though larger industrial units are known to be in operation [53, 95], no information concerning them has been released. The only experimental evidence for a bed larger than 61 cm diameter comes from a test carried out by Peterson [180] which showed the measured value of U_{ms} for a 91.4 cm (3 ft) diameter × 119 cm (~4 ft) deep wheat bed, with a 60° conical base and a 15 cm diameter air orifice, to be nearly double that predicted by Eq. (2.38). An examination of the form of Eq. (2.38) suggests that the above finding, even though based on a single result, may well indicate the general trend with increasing column diameter. It is seen from Fig. 2.9 that for geometrically similar beds (i.e., fixed values of H/D_c and D_c/D_i) of a given material, Eq. (2.38) predicts a continually decreasing value of U_{ms} and therefore of the jet velocity through the orifice, with increasing column diameter. Hence, for a sufficiently large value of D_c, the equation would eventually predict an orifice velocity lower than the free fall velocity of the particles. This is an absurd prediction since, unless the orifice velocity is well above the free fall velocity, particles will not be entrained as a dilute phase by the entering gas and spouting action cannot then possibly occur. For the system of Fig. 2.9, such an obvious breakdown of the equation does not occur until D_c becomes as large as 150 cm, but an increasing discrepancy between the predicted and experimental values would be expected to arise at some smaller value of D_c. The suggestion of such a trend is discernible even at $D_c = 61$ cm, since Thorley *et al.* [228] have noted that their experimental U_{ms} values for wheat beds of this size were consistently higher than prediction from Eq. (2.38), though only by about 10%. Becker's equation [Eq. (2.42)], too, predicts a continual decrease in U_{ms} with increasing D_c, as illustrated in Fig. 2.9, which is at least partly explained by the fact that his companion equation for calculating U_m [Eq. (2.35)] neglects the dependence of U_m on column diameter, as pointed out in Section A. The decrease predicted by Becker's method is, however, more gradual than that predicted by Eq. (2.38), and consequently U_{ms} for $D_c = 91.4$ cm given by Eq. (2.42) is considerably higher and closer to the experimental curve than that from Eq. (2.38). Becker's equation might therefore be expected to give closer prediction for columns somewhat larger than 61 cm diameter than the Mathur–Gishler equation, although if U_{ms} really does increase with D_c in large columns, as suggested by Peterson's result, Becker's equation too would begin to underestimate U_{ms} for beds much larger than 61 cm diameter.

3 | *Flow Pattern of Gas*

Detailed knowledge of the gas flow pattern in a spouted bed is important in ascertaining the effectiveness of gas–solids contact. A considerable volume of experimental data on flow distribution between the spout and the annulus has been reported. The effect of bed geometry parameters has been thoroughly studied, but mostly for a single bed material (wheat) and fluid (air). Attempts to give a generalized description of flow distribution, both theoretical and empirical, have also been made.

3.1 EXPERIMENTAL FINDINGS

Gas flow profiles have been determined by two methods:

(1) From measured longitudinal static pressure gradients at the column wall;
(2) By pitot tube measurements of local gas velocities.

The former method was used by Mathur, Thorley, and co-workers [137, 227, 228], who experimentally determined the relationship between superficial velocity and pressure drop for the bed particles in the static loose packed condition. With the assumption that the spouted bed annulus is similar in voidage to a loosely packed bed, measured vertical static pressure drops along the column wall during spouting could be directly related to superficial velocities and, when the cross-sectional area of the annulus as a function of bed level was known from semicircular column measurements, to volumetric flow rates in the annulus at each level. Flow through the spout was then obtained by difference. All observations were made in the region above the cone, where the pressure across a horizontal section was found to be almost uniform. As pointed out by Thorley *et al.* [227], the assumption that the annular solids behave hydrodynamically as a loose packed static bed is only an approximation. While the assumption was supported by pressure drops measured for the same solids in a moving bed, which is known to display the same voidage as a loose packed stationary bed [89] and which visually resembles the spouted bed annulus, the particles near the bottom were in fact observed to be somewhat more tightly packed than those at the top of the annulus. This nonhomogeneity would introduce some error in this method of estimating the gas distribution, mainly at the extremities of the cylindrical portion of the bed.

The data of Fig. 1.6 were obtained by this method. These data show that a substantial proportion of the gas cross-flows into the annulus within a short distance from the inlet, the final proportion being higher for larger columns. Further data, obtained in 61 cm diameter wheat beds, have been reported [227] for different cone angles (45–85°), inlet diameters (51–102 mm), bed depths (91–185 cm), and air flows (up to $1.3U_{ms}$). These are

TABLE 3.1   Percentage of Total Air Passing through Annulus at Bed Level $z/H = 0.5$ [a]

Variable	Value	Air flow through annulus (%)	Variable	Value	Air flow through annulus (%)
$D_i = 5.08$ cm, $H = 183$ cm, $U_s = 1.1U_{ms}$			$\theta = 60°$, $D_i = 5.08$ cm, $H = 152$ cm		
Cone angle	45°	61	Gas flow rate	$1.1U_{ms}$	48
	60°	55		$1.2U_{ms}$	42
	85°	48		$1.3U_{ms}$	37
$\theta = 60°$, $H = 183$ cm, $U_s = 1.1U_{ms}$			$\theta = 60°$, $D_i = 5.08$ cm, $U_s = 1.1U_{ms}$		
Orifice diameter	5.08 cm (2 in.)	54	Bed depth	91 cm (36 in.)	79
	7.62 cm (3 in.)	64		122 cm (48 in.)	67
	10.16 cm (4 in.)	68		157 cm (62 in.)	58
				185 cm (73 in.)	56

[a] Wheat beds, column diameter = 61 cm (24 in.); data of Thorley *et al.* [227].

illustrated by the results for an arbitrary bed level of $z/H = 0.5$ in Table 3.1, the trends of variation in the percentage flow figures with respect to each variable being similar at other fractional bed levels. It is seen that cross flow of gas into the annulus is favored by the use of a small cone angle, a large orifice size, a shallow bed, and a low gas flow rate. At higher flow rates, the additional gas apparently passes through the spout without spreading into the annulus [228, 15]. The observed trend with bed depth, namely, larger proportion of air flowing through the annulus for shallower beds, seems intuitively incorrect. It is conceivable that the nonhomogeneity in packing density mentioned above introduced significant errors in the results for the shallow beds, which extended only a short distance above the top of the cone (height of 60° cone = 44.5 cm).

The pitot tube method has been used by Becker [15], Mamuro and Hattori [131], and most recently by van Velzen *et al.* [244]. Typical deep bed ($H \sim H_m$) velocity profiles obtained in a 15.2 cm diameter column using wheat are shown in Fig. 3.1. Radial gas velocity profiles for shallower beds are similar except that the trough between the spout and the wall

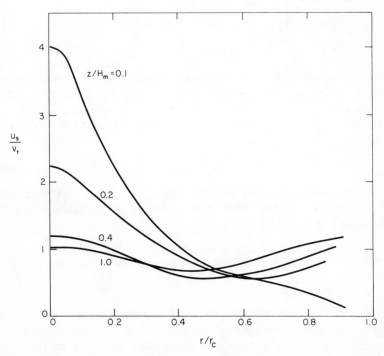

Fig. 3.1. Radial gas velocity profiles for wheat beds spouted in a 15.2 cm diameter flat-base column [15].

tends to be more pronounced [15, 131]. The development of a crest near
the wall could possibly be attributed to the higher than average local
porosity normally occurring at the walls of a packed bed [197], although
the absolute accuracy of pitot tube measurements in the annulus is open to
question because of the low gas velocities involved and the possible disturbance of local porosity by the introduction of the pitot tube. Thus, the
anomalously high value of gas velocity in the annulus which Becker reported has been queried [142], while Mamuro and Hattori were unable to
reproduce their total gas flows by integrating their radial gas velocity
profiles. They therefore dismissed their annulus data and used only the
local gas velocity results obtained in the spout to calculate the gas distribution between spout and annulus. Because of the high voidage prevailing
throughout most of the spout, pitot tube measurements in the spout should
be much less subject to the above errors, except possibly in the lower voidage
region near the top of the bed.

The vertical profiles of interstitial air velocity measured at the spout axis
for deep beds ($H \sim H_m$) of wheat spouted in 15.2 cm and 23 cm diameter
columns were empirically approximated by Becker as

$$u_s/v_t = 0.45/(z/H) \qquad (0.1 \lesssim z/H \lesssim 0.45, \quad H \sim H_m) \qquad (3.1a)$$

and

$$u_s/v_t \sim 1 \qquad (0.45 \lesssim z/H \lesssim 1, \quad H \sim H_m) \qquad (3.1b)$$

v_t being the terminal free fall velocity of a particle. The roughly constant
interstitial velocity u_s in the upper region of the bed [Fig. 3.1 and Eq. (3.1b)]
implies a decreasing superficial spout velocity $\overline{U}_s$ with increasing bed level
due to the marked decrease of porosity ϵ_s with bed level (see Fig. 5.3) in
this region:

$$\overline{U}_s = u_s \epsilon_s \qquad (3.2)$$

However, according to Richardson and Zaki [194], for the relatively high
particle Reynolds numbers prevailing in the spout, if negligible particle
acceleration is assumed,

$$u_s - v = v_t \epsilon_s^{2.39}/\epsilon_s \qquad (3.3)$$

or

$$u_s = v + v_t \epsilon_s^{1.39} \qquad (3.3a)$$

where v is the upward particle velocity. Equation (3.3a) would appear to
be applicable to the upper region of the bed because, as shown in the following section, the initial accelerating momentum imparted to the spout
particles by the high jet velocity at the inlet orifice has been dissipated by
the time the particles get to the upper region. Since in this region ($z/H > 0.45$)

both v (Figs. 4.2 and 4.3) and ϵ_s (Fig. 5.3) are decreasing with z, Eq. (3.3a) implies that u_s must also decrease with z, instead of remaining constant. The anomaly disappears when it is realized that kinetic energy and hence velocity additional to that given by Eq. (3.3a) must be supplied by the gas to accelerate cross-flowing particles from the annulus to the particle velocities which prevail in the spout at various levels.

A different approach to correlating the axial decay of spout gas velocity u_s was used by van Velzen *et al.* [244], who selected the momentum of the entering gas jet as the correlating parameter. Experimental data covering a range of particle properties ($d_p = 0.5–1.3$ mm, $\rho_s = 1.0–4.4$ Mg/m³), nozzle sizes, and gas flow rates obeyed an empirical equation developed on this basis, vindicating the choice of jet momentum as the single parameter to describe the combined effect of the variables studied. Their equation itself, however, is valid only for the particular bed geometry used in the experiments ($D_c = 12.5$ cm diameter, $H = 20$ cm, $\theta = 31°$), and for $z \leq 15$ cm.

Measuring the gas distribution between spout and annulus in the lower conical part of the bed has proved to be difficult. Abdelrazek [1] used a probe to determine radial static pressure profiles at different levels in a small bed of glass spheres ($D_c = 7.6$ cm, $H = 16.0$ cm, $\theta = 60°$, $d_p = 0.643$ mm). Radial pressure gradients in the upper cylindrical section were found to be insignificant but in the lower part of the bed and particularly within the conical section, static pressures recorded at the spout axis were somewhat lower than at the spout–annulus interface. From the latter observation, Abdelrazek was inclined to infer that a reverse flow of gas from the annulus into the spout must be occurring in the lower part of the bed. He did, however, realize that the existence of a small adverse pressure gradient does not rule out the possibility of gas flow in the opposite direction, i.e., from the spout to the annulus, since at least part of the radially directed gas kinetic energy in the spout may be recovered as pressure in the annulus. Evidence of strong recirculation of gas in the lower part of the bed—up to a vertical distance of 5–7 cm from the inlet in 12.5 cm diameter × 20 cm deep beds—has also been reported by van Velzen *et al.*, again on the basis of static pressures measured in the annulus. The vertical gradients in the lower part of the annulus were often found to be negative, but these investigators failed to consider the possibility that this finding might be due to a pressure recovery effect similar to that mentioned above rather than to reverse flow of gas. Their reported values of downward velocities near the bottom of the annulus (up to 7.6 m/sec) seem too high to be plausible when compared with qualitative gas tracer observations made in our laboratories. These results tend to support the view that static pressure measurements might not give a reliable indication of gas flow pattern in the vicinity of the inlet orifice, where the gas jet undergoes sudden expansion. Some form of quantitative tracer technique would seem necessary for investigating the movement of gas in this region. The flow pattern here could well turn out to be more complex than the generally accepted picture of the gas simply flaring out from the spout into the annulus as it travels upward, which is based on observations made primarily in the cylindrical part of the bed above the conical base.

3.2 PREDICTION OF FLOW PATTERN

Mamuro and Hattori [131] have theoretically derived an equation which enables estimation of gas distribution between the annulus and the spout at various levels in the bed. The derivation starts with a force balance on a differential height dz of the annulus (see Fig. 3.2), neglecting friction at both vertical boundaries of the annulus:

$$-dP_b = (\rho_s - \rho_f)(1 - \epsilon_a)g\,dz - (-dP) \tag{3.4}$$

where P_b is the net downward solids force per unit cross-sectional area of the annulus. Since the interface between the spout and the annulus is maintained steady, the kinetic pressure due to cross flow of gas from the spout to the annulus is presumed to be in balance with the radial component of the annular bed pressure. Assuming, by analogy with Janssen's assumption for powders in hoppers [32, p. 70], that the vertical component of this pressure is proportional to the radial component, one can write

$$P_b = k\rho_f U_r^2/2 \tag{3.5}$$

U_r being the superficial radial gas velocity across the spout–annulus interface. Differentiating with respect to z, we have

$$dP_b/dz = k\rho_f U_r\,dU_r/dz \tag{3.6}$$

The annulus is taken to be a uniformly packed bed in which the solids velocity can be neglected and in which the relatively slow upward movement of gas is assumed to follow Darcy's law:

$$-dP/dz = KU_a \tag{3.7}$$

Combination of Eqs. (3.4), (3.6), and (3.7) leads to

$$k\rho_f U_r(dU_r/dz) + (\rho_s - \rho_f)(1 - \epsilon_a)g = KU_a \tag{3.8}$$

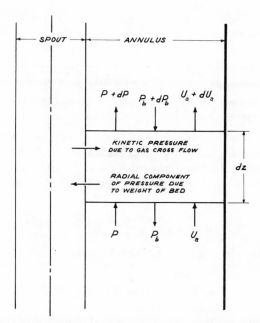

Fig. 3.2. Model for forces acting in the annulus [131].

Differentiating with respect to z, we obtain

$$k\rho_f \frac{d}{dz}\left(U_r \frac{dU_r}{dz}\right) = K \frac{dU_a}{dz} \tag{3.9}$$

From a gas balance over the height dz, neglecting any change in spout diameter with z,

$$\pi D_s U_r = A_a \, dU_a/dz \tag{3.10}$$

Combination and rearrangement of Eqs. (3.9) and (3.10) yields

$$\frac{d}{dz}\left(U_r \frac{dU_r}{dz}\right) = \frac{K\pi D_s}{k\rho_f A_a} U_r \tag{3.11}$$

the general solution of which is

$$U_r = C_1(z + C_2)^2 \tag{3.12}$$

Substituting for U_r in Eq. (3.10) according to Eq. (3.12), we have

$$dU_a/dz = (\pi D_s/A_a)C_1(z + C_2)^2 \tag{3.13}$$

At the base of the bed, the boundary condition is $z = 0$, $U_a = 0$. Therefore integration of Eq. (3.13) results in

$$U_a = \frac{\pi D_s C_1}{A_a} \int_0^z C_1(z + C_2)^2 = \frac{\pi D_s C_1}{3A_a}[(z + C_2)^3 - C_2^3] \quad (3.14)$$

Assuming that the gas velocity at the top of the annulus is just sufficient to fluidize the solids, i.e., that the bed is at its maximum spoutable depth H_m, then the relevant boundary conditions at the top of the bed, where $z = H = H_m$, are $dP_b/dz = 0$ [which makes Eq. (3.4) equivalent to Eq. (2.13) for minimum fluidization], and $U_a = U_{mf}$. After substitution of the first condition into Eq. (3.6), it follows from Eq. (3.12) that

$$\left(U_r \frac{dU_r}{dz}\right)_{z=Hm} = 0 = 2C_1^2(H_m + C_2)^3 \quad (3.15)$$

Therefore, $C_2 = -H_m$, so that Eq. (3.14) becomes

$$U_a = \frac{\pi D_s C_1}{3A_a}[H_m^3 - (H_m - z)^3] \quad (3.16)$$

From the second condition, Eq. (3.16) applied to the top of the bed becomes

$$U_{mf} = \frac{\pi D_s C_1}{3A_a} H_m^3 \quad (3.17)$$

Dividing Eq. (3.16) by Eq. (3.17) yields

$$U_a/U_{mf} = 1 - (1 - (z/H_m))^3 \quad (3.18)$$

For beds below the maximum spoutable, Mamuro and Hattori arbitrarily modified Eq. (3.18) to

$$U_a/U_{aH} = 1 - (1 - (z/H))^3 \quad (3.19)$$

For testing the validity of Eq. (3.19), Mamuro and Hattori made use of spout pitot tube data—their own as well as Becker's—integrating the velocity over the cross section of the spout. Most of these data were for wheat beds of different depths spouted in a 15 cm diameter column, although one set of results for soma sand in a 15 cm column and another set for wheat in a 22.8 cm column were also included. The agreement they obtained with Eq. (3.19) was only fair. Moreover, the reliability of the values of U_a/U_{aH} calculated by Mamuro and Hattori from pitot tube measurements is questionable, because of the somewhat arbitrary procedure adopted for estimating the spout voidage profile from reported data [137] obtained under quite different conditions. The voidage profile was required for converting interstitial velocities into superficial velocities. The static pressure method used by Mathur, Thorley, *et al.* [method (1) of Section 3.1], on the other hand, gives the superficial velocity in the annulus directly. These data, which have the further advantage of covering a variation in column size, therefore provide a better means for testing the validity of Eq. (3.19). The comparison, shown in Fig. 3.3, does lend a measure of support to the theoretical equation, but some of the results for the larger columns show considerable deviations. This is perhaps not surprising, considering that bed depths used in the larger columns were generally much smaller than the maximum spoutable depths and that the theoretical analysis of Mamuro and Hattori leads only to Eq. (3.18) (for $H = H_m$), its extension to Eq. (3.19) (for $H < H_m$) being arbitrary.

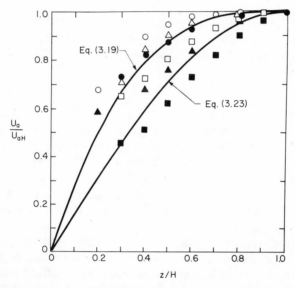

Fig. 3.3. Vertical profile of air distribution. Note that $H \sim H_m$ only in the case of the 15.2 cm column data.

	D_c (cm)	D_i (cm)	θ (deg)	H (cm)	Ref
△	15.2	1.27	85	64	[137]
○	30.5	5.08	85	114	[138]
■	61.0	5.08	85	183	[228]
▲	61.0	7.62	60	183	[228]
□	61.0	10.16	60	183	[228]
●	61.0	10.16	85	183	[227]

Table 3.2   Experimentally Determined Values of U_{aH}/U_s for Air–Wheat

Source	D_c (cm)	D_i (cm)	θ (deg)	H (cm)	U_{aH} (m/sec)	U_s (m/sec)	U_{aH}/U_s
Mathur and Gishler [137]	15.2	1.27	85	64	0.81	1.13	0.72
Mathur and Gishler [138]	30.5	5.08	85	114	0.82	0.91	0.90
Thorley *et al.* [228]	61.0	5.08	45	183	0.57	0.62	0.93
		5.08	60	91	0.38	0.38	1.0
		5.08	60	122	0.46	0.46	1.0
		5.08	60	152	0.51	0.55	0.93
		5.08	60	183	0.56	0.60	0.93
		7.62	60	183	0.62	0.69	0.91
		10.16	60	183	0.65	0.73	0.89
		5.08	85	183	0.51	0.55	0.92
		5.08	85	183	0.53	0.59	0.90
		5.08	85	183	0.52	0.65	0.80

It should be noted that the ratio U_a/U_{aH} obtained from Eq. (3.19) must be multiplied by $(U_{aH}/U_s)(A_a/A_c)$ in order to arrive at the gas fraction flowing through the annulus at any level. Values of U_{aH}/U_s estimated from available data are recorded in Table 3.2. Although variations in the cross-sectional area of the spout play a small role in determining variations in gas distribution from one set of conditions to another, in the main the degree to which the ratio U_{aH}/U_s approaches unity is a measure of the extent to which the gas has spread out from the spout into the annulus by the time it has reached the bed surface. The data for the 61 cm diameter column in Table 3.2 are consistent with the trends in gas distribution at the halfway level in the bed, illustrated in Table 3.1. More important, however, is the effect of column diameter revealed by a comparison of U_{aH}/U_s for the three different column sizes. The extent to which the gas has flared out into the annulus during its travel through the bed is seen to increase substantially with increasing column diameter, irrespective of the bed depth, cone angle, or orifice diameter used. Abdelrazek's results for beds of glass beads in smaller columns are in line with the above trend, the values of U_{aH}/U_s for his 7.6 cm and 10.2 cm diameter columns being in the region of 0.5.

The data in Table 3.2 also show that the value of U_{aH} itself does indeed equal U_{mf} when $H \sim H_m$ (or even $0.7H_m$), as was assumed by Mamuro and Hattori. Their suggestion that for bed depths considerably smaller than H_m, the ratio U_{aH}/U_{mf} increases with H/H_m (in line with Becker's similarity principle [15]) is also supported by the data of Table 3.2 (plotted in Fig. 3.4), but these data, which are more extensive and reliable than those of Mamuro and Hattori, show the relationship between U_{aH}/U_{mf} and H/H_m to be roughly parabolic rather than linear. In computing values of H/H_m, experimental results for H_m reported by Mathur and Gishler [137, 138] were used for the 15.2 cm and 30.5 cm diameter columns, while values estimated by Eq. (6.5) [129] were taken for the 61 cm column. The lack of agreement with the data of Mamuro and Hattori (also shown in Fig. 3.4) is perhaps not surprising in view of the difficulties, already discussed, in obtaining annulus air velocities from spout pitot tube measurements. Nevertheless, their results show a more reasonable trend with respect to varying bed depth, namely, larger proportion of gas through annulus for deeper beds, than that shown in Table 3.1.

It should be emphasized that the empirical generalization represented by Fig. 3.4 is based on experimental results for wheat only, but the parameters chosen do include the effect of solid and fluid properties through both U_{mf} and H_m, and of column geometry variables (with the exception of cone angle) through H_m. Experimental data with materials other than wheat are, however, necessary to test the generality of the relationship in Fig. 3.4.

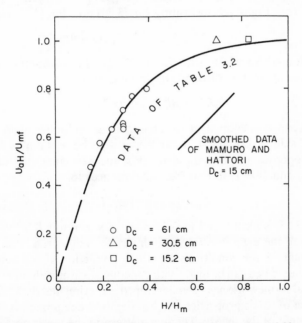

Fig. 3.4. Variation of U_{aH}/U_{mf} with H/H_m for air–wheat system, $U_{mf} = 0.82$ m/sec (exptl). With reference to the smoothed data of Mamuro and Hattori [131], note that $H = $ total bed depth measured from gas inlet $\neq$ bed depth above cone as taken by them.

A proven correlation of this type would be of considerable value, since in conjunction with Eq. (3.19) and the available equations for estimating U_{mf}, U_{ms}, and H_m, it would enable the gas flow distribution in any given bed to be predicted, without experimentation.

The limitation with respect to solids mentioned above applies not only to the plot of Fig. 3.4 but also to the verification of Eq. (3.19). Although the implication that longitudinal flow distribution between the spout and the annulus is independent of solids properties is implicit in the derivation of Eq. (3.19), the experimental data supporting the equation are for wheat except for a single run by Mamuro and Hattori [131] on soma sand ($d_p = 1.18$ mm, $\rho_s = 2.61$ Mg/m^3). Thus experimental verification on this point is again necessary.

An alternative approach to predicting gas distribution is that of Lefroy and Davidson [114]. Equation (2.24), which as written applies only for the condition $H = H_m$, is generalized to any spoutable bed depth by writing instead

$$-\Delta P_s = (2BH/\pi)(\rho_s - \rho_f)(1 - \epsilon_{mf})g \tag{3.20}$$

where the factor B (≤ 1) is the ratio $(\Delta P_s/H)/[(\Delta P_s)_{max}/H_m]$ and can be estimated from Fig. 2.7. Substituting Eq. (3.20) into Eq. (2.23) results in

$$-dP/dz = B(\rho_s - \rho_f)(1 - \epsilon_{mf})g \sin(\pi z/2H) \tag{3.21}$$

Combination of Eq. (3.21) with Eq. (2.17) on the assumption of Darcy's law (i.e., $n = 1$) for flow through the annulus yields

$$U_a = BU_{mf} \sin(\pi z/2H) \tag{3.22}$$

Since the term B is equivalent to the empirical multiplier of H/H_m on the right-hand side of Eq. (2.27), it can be seen to be a complex function of particle properties and bed geometry. For the particular case of $H = H_m$, $U_{mf} = U_{aH}$ and $B = 1$, so that Eq. (3.22) reduces to

$$U_a/U_{aH} = \sin(\pi z/2H) \tag{3.23}$$

Equation (3.23) is plotted in Fig. 3.3, where it is seen to fall below Eq. (3.19) but to match the experimental data almost as well as the latter equation.

To summarize, the conditions for which prediction of gas flow distribution in a spouted bed can be made with confidence are quite limited. Although a theoretical framework has been developed, experimental data, particularly on the effect of solids properties and bed depth, are required for verification and refinement of the theory. The flow pattern in the lowermost part of the bed, which could be important, remains largely unexplored. Here, there is no clear definition of the spout shape (see Chapter 5), there is a definite suction effect [13], and there is pronounced gas pulsation [248]. Investigation of the movement of gas in this region, therefore, presents a challenging problem, both experimentally and theoretically.

3.3 RESIDENCE TIME DISTRIBUTION

So far, we have looked upon the gas flow pattern as essentially the bulk distribution of gas between the spout and the annulus, disregarding smaller-scale aspects of flow pattern, such as axial mixing and radial velocity gradients in the individual channels. The importance of these effects is best determined by comparing the actual residence time distribution of gas measured by tracer experiments against the expected distribution from an idealized bulk flow model. A simple two-region model of this type, which postulates plug flow of gas in the two channels of a spouted bed with cross flow occurring from the spout to the annulus, is presented in Chapter 10 for a spouted bed chemical reactor, and can be applied for predicting the residence time distribution of gas.

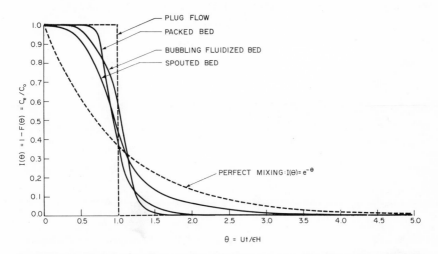

Fig. 3.5. Output response to step reduction in gas tracer input [121]. Column: 15 cm diameter, semicircular cross section; bed solids: $-10 +30$ mesh Ottawa sand; settled bed depth: 46 cm. Gas: air, tracer: helium ($t \leq 0$, $C_i = C_0 = C_e$; $t \geq 0$, $C_i = 0$, $C_e \leq C_0$). Spouted bed: $D_i = 2.1$ cm, cone angle = $60°$, $U = 31$ cm/sec. Fixrd packed bed: uniform gas distributor, $U = 30$ cm/sec. Bubbling fluidized bed: uniform gas distributor, $U = 61$ cm/sec, $H_f = 76$ cm.

At the relatively high particle Reynolds numbers occurring in both spout and annulus (Re > 10), the degree of axial mixing in the individual channels is likely to be small [57]. The large difference in flow velocities in the two channels would nevertheless cause the residence time distribution in the bed as a whole to deviate substantially from plug flow behavior. The radially nonuniform flow within both spout and annulus, especially in the lower part of the bed, and any small-scale recirculation of gas in the entrance region would accentuate this deviation.

The only available stimulus–response data are shown in Fig. 3.5, where the measured outlet response to an imposed step change in the inlet gas concentration is plotted for essentially the same bed of solids in the packed, spouted, and fluidized states. To obtain a well fluidized bed, the gas velocity and hence Reynolds number had to be twice as large as the values for the packed and spouted beds. The smallest deviation from plug flow behavior is observed for the fixed packed bed, which is commonly represented by the axial-dispersed plug flow model [120]. The largest deviation is observed for the spouted bed, the response curve of which is nevertheless still far removed from that of perfect mixing. The nonuniform flow in a fluidized bed obviously increases the axial dispersion above that for a packed bed, but apparently not to the same extent as in a spouted bed, at least in the present instance. Comparison of the spouted bed data with prediction from the model mentioned above has not yet been made.

4 | *Flow Pattern of Solids*

The solids movement in a spouted bed is initiated by the interaction between the particles and the high velocity gas jet, so that particle flow in the spout region has a strong influence on the entire solids flow pattern. While the motion of individual particles in the spout and in the annulus and the gross mixing behavior of solids are all interdependent, it is nevertheless convenient to discuss these three aspects of solids flow pattern separately.

4.1 PARTICLE MOTION IN SPOUT

A particle starting from the base of the bed first accelerates from rest to a peak velocity, and then decelerates until it again reaches zero velocity at the top of the fountain, where it reverses its direction of movement. In addition, a radial profile of longitudinal velocity exists at any particular level, the velocity at the axis being higher than at other radial positions. After summarizing the two techniques which have been used for measuring particle velocity in the spout, we will first consider the longitudinal movement of the particles, neglecting until later the radial profile.

A. Measurement Techniques

1. Direct Observation

It is not possible to observe the particle motion in the spout in the usual orifice-centered, three-dimensional, spouted column since the spout, being surrounded by the annular solids, is not open to view. If, however, the bed is contained in a column of semicircular cross section, with matching conical base and inlet orifice, a longitudinal section of the bed through the axis becomes visible against the flat transparent face of the column. Particle velocities in the spout can then be measured by high speed motion pictures.

The question as to whether or not a half-sectional column truly represents the behavior of a spouted bed in a whole column has been experimentally investigated by obtaining comparative data on various aspects of the behavior in the two types of columns. While Mathur and Gishler [137], who first used a half column, had expressed some reservation on this point, subsequent evidence [152, 114] has shown that any differences which exist are not significant for most purposes. A possible difference with respect to the detailed spout contour (see Chapter 5), especially near the gas entrance, is an exception, as yet uninvestigated, which may be of major importance to the quantitative analysis of spouting dynamics.

2. Piezoelectric Technique

This technique enables the spout particle velocities to be measured in a whole column. The method was developed by Gorshtein and Soroko [83] and subsequently used for spout velocity measurements by other Soviet workers. It is based on the piezoelectric effect, namely, that crystals of certain materials, such as piezoquartz, barium titanate, and Rochelle salt, when subjected to a short-term mechanical shock, generate an electric charge on the surface of the crystal, the magnitude of the charge being a function of the force with which the crystal is hit. A piezocrystal of a size similar to that of the bed particles is used as the sensing element, and is mounted to serve as a probe for measuring velocity profiles in the spout. This is connected to an amplifying device and then to an oscilloscope which records the charge generated. Prior calibration of the instrument with the same solid particles is necessary to convert the electric charge data into particle velocities. The lower limit of sensitivity of the technique has been quoted as 10–20 cm/sec for particles of 2 mm diameter [83].

B. Longitudinal Profile

1. *Theoretical Analysis*

Based on particle force balance   Thorley *et al.* attempted to predict the longitudinal particle velocity profile in the spout on the basis of a force balance analysis, making some changes in assumptions between their preliminary report [227] and their subsequent paper [228]. The model presented below comes quite close to the earlier approach of Thorley *et al.,* the main modifications being the incorporation of more realistic gas flow behavior, and the allowance made for the shape of the spout in the lower part of the bed. Their earlier model was selected for making the above modifications,

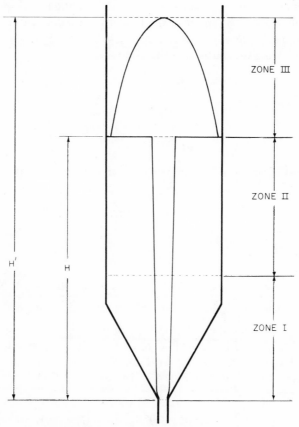

Fig. 4.1.   Schematic for particle force balance model.

since the neglect of particle cross flow in their later model appeared to be a gross oversimplification.

Consider the spout as being made up of three successive zones, corresponding to the regions of initial particle acceleration, subsequent particle deceleration within the bed proper, and final deceleration in the fountain, respectively. The three zones are illustrated in Fig. 4.1. A general force balance is written for the spout particles and its solution is carried out by introducing specific simplifications and boundary conditions for each zone.

The only accelerating force on the spout particles is the frictional drag of the ascending fluid, while the main decelerating force is that of gravity. The latter force is reinforced by cross-flowing particles, which enter the spout at zero velocity and hence tend to lower the average velocity of the rising particles. It is assumed that there is instantaneous equalization of particle velocities in the spout at any level, the cumulative number of rising particles at a given level (i.e., after time t; $t = 0$ at $z = 0$) being n_z. The force balance following the particles is therefore

$$\tfrac{1}{6}\pi d_v{}^3 \rho_s \, d(n_z v)/dt = (n_z \tfrac{1}{2}\rho_f(u_s - v)^2 \tfrac{1}{4}\pi d_p{}^2 C_D) - (n_z \tfrac{1}{6}\pi d_v{}^3(\rho_s - \rho_f)g) \quad (4.1)$$

Using the identity

$$\frac{d(n_z v)}{dt} \equiv v\frac{dn_z}{dt} + n_z\frac{dv}{dt} \tag{4.2}$$

and converting from a Lagrangian to an Eulerian basis by means of the relationship

$$dt = dz/v \tag{4.3}$$

reduces Eq. (4.1) to

$$\frac{v^2}{n_z}\frac{dn_z}{dz} + \frac{v\,dv}{dz} = \frac{3\rho_f(u_s - v)^2\,d_p{}^2 C_D}{4d_v{}^3\rho_s} - \frac{(\rho_s - \rho_f)g}{\rho_s} \tag{4.4}$$

The term dn_z/dz represents that change of particle concentration with bed level which is caused by cross flow of particles into the spout.

In zone I, solids cross flow from the annulus is assumed to be concentrated at the base of the bed (see Fig. 4.7), i.e., $dn_z/dz = 0$. Therefore, Eq. (4.4) simplifies to

$$\frac{v\,dv}{dz} = \frac{3\rho_f(u_s - v)^2\,d_p{}^2 C_D}{4d_v{}^3\rho_s} - \frac{(\rho_s - \rho_f)g}{\rho_s} \tag{4.5}$$

The same simplification was made by Thorley *et al.*, who initially [227] assumed in addition that $u_s = u_i$ throughout zone I, whereby they were able to integrate Eq. (4.5) analytically. Later [228], however, they realized that u_s was a strong function of z in the gas entrance region, and therefore assumed it to vary linearly with z, for both zone I and zone II. It is proposed here instead that u_s for zone I be determined by the relationships of the previous chapter, so that Eq. (4.5) must be integrated numerically. The lower boundary condition is $u_s = u_i$ and $v = 0$ at $z = 0$, while the upper boundary is located by intersection with Eq. (4.9).

It is further proposed that by the time the gas has reached zone II, its velocity has decreased sufficiently that as an approximation, the drag term may be neglected relative to the other terms in Eq. (4.4), which then simplifies to

$$\frac{v^2}{n_z}\frac{dn_z}{d_z} + \frac{v\,dv}{dz} = -\frac{(\rho_s - \rho_f)g}{\rho_s} \tag{4.6}$$

Assuming after Thorley *et al.* [227] that in this zone n_z is directly proportional to z, an approximation which is justified only by its simplicity and by the rough linearity of particle concentration with bed level (Fig. 5.3) in this zone of nearly constant u_s (Fig. 3.1), then Eq. (4.6) reduces to

$$\frac{v^2}{z}\frac{d(v^2)}{2\,dz} + \frac{(\rho_s - \rho_f)g}{\rho_s} = 0 \tag{4.7}$$

For $z \geqq H$, $dn_z/dz = 0$ and since for $z = H'$, $v = 0$, it follows from Eq. (4.6) that

$$v_H{}^2 = \frac{2g(\rho_s - \rho_f)(H' - H)}{\rho_s} \tag{4.8}$$

Equation (4.8) is the upper boundary condition for zone II. Equation (4.7) can then be integrated analytically using this boundary condition, the final result being

$$v^2 = \frac{2g(\rho_s - \rho_f)}{\rho_s}\left(\frac{3H'H^2 - 2H^3 - z^3}{3z^2}\right) \tag{4.9}$$

In zone III, the fountain, particle cross flow into the spout is entirely absent (i.e., $dn_z/dz = 0$) and Eq. (4.4) further simplifies to

$$\frac{v\,dv}{dz} = -\left(\frac{\rho_s - \rho_f}{\rho_s}\right)g \tag{4.10}$$

The upper boundary condition is $v = 0$ at $z = H'$, so that integration of Eq. (4.10) yields

$$v^2 = \frac{2g(\rho_s - \rho_f)}{\rho_s}(H' - z) \tag{4.11}$$

Equation (4.11) matches Eq. (4.9) at $z = H$.

Based on mass and momentum balances   A more fundamental approach was attempted by Lefroy and Davidson [114], who wrote differential mass and momentum balances for both the gas and the solids flow in the spout. The mass balance for the gas is

$$\tfrac{1}{4}\pi D_s^2\,(d(\epsilon_s u_s)/dz) + \pi D_s U_r = 0 \tag{4.12}$$

while that for the solids is

$$\rho_s\tfrac{1}{4}\pi D_s^2\,(d[(1 - \epsilon_s)v]/dz) - (dW/dz) = 0 \tag{4.13}$$

The corresponding momentum balances are

$$\rho_f\frac{d(\epsilon_s u_s^2)}{dz} = -\epsilon_a\frac{dP}{dz} - \beta(u_s - v)^2 \tag{4.14}$$

and

$$\rho_s\frac{d[(1 - \epsilon_s)v^2]}{dz} = -(1 - \epsilon_s)\frac{dP}{dz} - (\rho_s - \rho_f)(1 - \epsilon_s)g + \beta(u_s - v)^2 \tag{4.15}$$

respectively. The gas–solid interaction factor β in the latter equation is evaluated from considerations of particulate fluidization ($v = 0$), for which Eq. (4.15), combined with the condition that pressure gradient equals the bouyed weight of the fluidized bed, reduces to

$$\beta u_s^2/\epsilon_s = (\rho_s - \rho_f)(1 - \epsilon_s)g \tag{4.16}$$

For particulate fluidization, u_s may be obtained from the Richardson–Zaki [194] equation, which at high Reynolds numbers, with its terminal velocity term evaluated in the Newton regime ($C_D = $ constant), reduces to

$$u_s = \left[\frac{4(\rho_s - \rho_f)gd_p}{3C_D\rho_f}\right]^{1/2}\epsilon_s^{1.39} \tag{4.17}$$

Combination of Eqs. (4.16) and (4.17) gives

$$\beta = \frac{3C_D\rho_f(1 - \epsilon_s)}{4d_p\epsilon_s^{1.78}} \tag{4.18}$$

Turning now to the gas balance, Eq. (4.12), and combining it with Eq. (3.10) for gas flow into the annulus, the result is

$$D_s^2 \frac{d(\epsilon_s u_s)}{dz} + (D_c^2 - D_s^2)\frac{dU_a}{dz} = 0 \qquad (4.19)$$

which can be obtained alternately by differentiating Eq. (4.22) with respect to z. Substituting into Eq. (4.19) according to Eq. (3.22) gives

$$-d(\epsilon_s u_s) = BU_{mf}((D_c^2/D_s^2) - 1)(\pi/2H)\cos(\pi z/2H) \qquad (4.20)$$

Integration of Eq. (4.20) using the lower boundary conditions, $\epsilon_s = 1.0$ and $u_s = u_i$ at $z = 0$, results in

$$u_i - \epsilon_s u_s = BU_{mf}((D_c^2/D_s^2) - 1)\sin(\pi z/2H) \qquad (4.21)$$

Numerical integration of Eqs. (4.18) and (4.19) in conjunction with Eqs. (3.21) and (4.21), along with the additional bottom condition, $v = 0$ at $z = 0$, yields a vertical profile of ϵ_s, u_s, and v. Equation (4.13) can then be used to compute the solids cross flow dW/dz.

2. Comparison of Theory with Experimental Data

Particle force balance theory In testing this theory against experimental data, the main difficulty which arises is obtaining an accurate knowledge of $u_s(z)$ in zone I. This not only requires some method for predicting how the gas divides itself between spout and annulus, which is provided at least approximately by Eq. (3.19) in conjunction with Fig. 3.4, but even more importantly it requires an accurate knowledge of spout diameter as it varies with bed level in the immediate vicinity of the gas inlet orifice [102a]. While an average spout diameter for the bed as a whole can be reasonably estimated from an empirical relationship such as Eq. (5.4), there is no accurate information available on the variation of D_s with z for the gas entrance region, i.e., the lower part of zone I. The spout can either contract or expand substantially and abruptly in this region under certain conditions, as discussed in Chapter 5.

Figures 4.2 and 4.3 show experimental data on spout velocities obtained photographically from a 61.0 cm diameter [228] and a 15.2 cm diameter [137] half column, respectively. In the larger column, D_s from Eq. (5.4) differed only a little from D_i and was therefore taken as the spout diameter in zone I. In the smaller column, however, D_s from Eq. (5.4) was more than three times D_i, and therefore the spout diameter in zone I was arbitrarily assigned a value equal to the geometric mean of D_s from Eq. (5.4) and D_i.

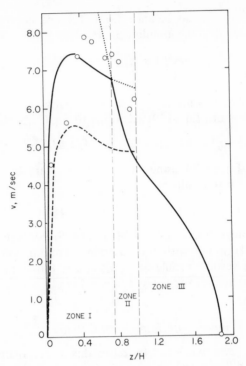

Fig. 4.2. Longitudinal profile of particle velocity in spout for 61 cm diameter × 122 cm wheat bed: $D_i = 10.2$ cm, $D_s = 8.1$ cm, $d_p = 3.2$ mm, $\rho_s = 1.376$ Mg/m^3, $U_s = 0.68$ m/sec, $U_{aH} = 0.95 U_s$, $C_D = 1.0$; $\bigcirc$ experimental values of Thorley *et al.* [228], —— from particle force balance theory, – – – from mass and momentum balance theory with B [in Eq. (4.21)] taken as 0.6.

Values of U_a/U_{aH} were estimated from Eq. (3.19) and U_{aH}/U_s from Table 3.2, from which the spout velocity u_s at any bed level could be evaluated from the gas balance equation,

$$U_s D_c^2 = \epsilon_s u_s D_s^2 + U_a(D_c^2 - D_s^2) \tag{4.22}$$

ϵ_s being assumed equal to unity in zone I. In the absence of accurate data on the characteristics of the wheat particles, the term d_v^3/d_p^2 was taken as the smaller dimension (diameter) of the wheat grains, which have been observed to align themselves with their longer dimension (axis) parallel to the flow [137], ignoring the small difference between the size of wheat used in experiments with the two column sizes (recorded in Table 2.4).

Given the approximations inherent in the theoretical model and the uncertainties involved in its numerical evaluation, the trends displayed by the model are in accord with the experimental trends, and even the absolute

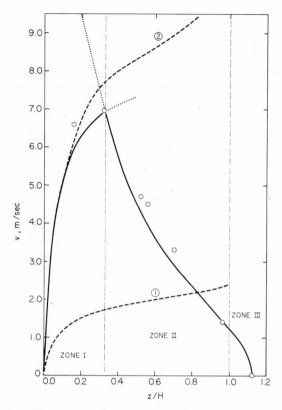

Fig. 4.3. Longitudinal profile of particle velocity in spout for 15.2 cm diameter × 63.5 cm wheat bed: $D_i = 1.25$ cm, $d_p = 3.2$ mm, $\rho_s = 1.376$ Mg/m³, $U_s = 1.13$ m/sec, $U_{aH} = 0.7U_s$, $C_D = 1.0$; ○ experimental values of Mathur and Gishler [137], ——— from particle force balance theory, with D_s taken as $(4.1 \times 1.25)^{1/2} = 2.26$ cm for all z/H, ――― from mass and momentum balance theory, with B taken as 0.95. Curve 1: $D_s = 4.1$ cm for all z/H; curve 2: $D_s = 1.25 + (4.1 - 1.25/0.33 (z/H)$ up to $z/H = 0.33$, and $D_s = 4.1$ cm for $0.33 \leq z/H \leq 1$.

agreement with experimental data in Figs. 4.2 and 4.3 is reasonable. The measurements recorded in these figures tended to be biased toward particle velocities at the spout axis [228], which, as will presently be seen, are about twice the average particle velocities. This bias in the experimental values seems a plausible explanation for the fact that almost all the points fall above the theoretical curves. Only in zone I, where the spout diameter had to be guessed at in evaluating the model, are there any experimental points which fall below the predicted curves.

Mass and momentum balance theory Predicted particle velocities according to this method are also shown in both Figs. 4.2 and 4.3. In Fig. 4.2, where the spout diameter for the whole column differed only a little from

D_i and was therefore taken as the value given by Eq. (5.4), the computed trend is credible, except for the flattening out of the profile toward the top of the bed. In Fig. 4.3, however, the determination of D_s was again a problem and was done in two alternate ways. The bottom curve was based on the value of D_s as given by Eq. (5.4) for the whole column, while the top curve was based on the assumption that D_s varied linearly from D_i at $z = 0$ to the value given by Eq. (5.4) at $z = H/3$ (i.e., in the region corresponding to zone I of the previous model), the latter value remaining constant for the upper two-thirds of the bed. Neither curve shows any deceleration zone, and the large difference between them points to the very great sensitivity of the Lefroy–Davidson method to the assumed spout contour. Thus an accurate knowledge of spout diameter and its variation with bed level, particularly near the jet entrance, appears to be essential for getting correct answers by the fundamental mass and momentum balance approach.

It should also be pointed out that all the above calculations, as well as those performed by Lefroy and Davidson [114] themselves, show values of u_s which continue to fall with increasing z all the way to $z = H$. This aspect

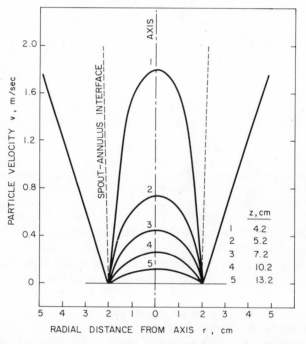

Fig. 4.4. Radial profiles of upward particle velocities in the spout, after Gorshtein and Mukhlenov [86]: Conical vessel, $\theta = 30°$, $D_i = 4.0$ cm, $H = 12.0$ cm; catalyst marbles, $d_p = 1.5$ mm, $u_i = 3.47$ m/sec.

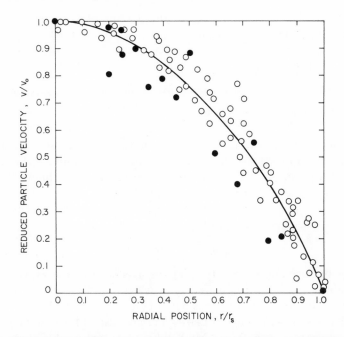

Fig. 4.5. Normalized plot of spout particle velocities. ○ Gorshtein and Mukhlenov [86]: Conical vessels, catalyst marbles, $d_p = 1.5$ mm; $\theta = 20°$–$60°$, $H/D_i = 1.5$–6.0, $z = 5$ cm and 10 cm. ● Lefroy and Davidson [114]: cylindrical vessel, $D_c = 30.5$ cm; polyethylene chips, $d_p = 3.5$ mm; $\theta = 180°$, $H/D_i = 39$, z not given.

of their theory too is at odds with experimental data, namely, those summarized by Eq. (3.1b).

3. *Other Work*

Using the piezoelectric technique, Mikhailik and Antanishin [153] measured particle velocities of millet, silica gel, and polystyrene ($d_p = 1.5$–7.0 mm) spouted in a 9.4 cm diameter conical-cylindrical column. The longitudinal profiles obtained were similar in shape to those of Figs. 4.2 and 4.3. For a given gas velocity, smaller particles reached a higher peak value and took a longer distance to do so than larger particles. Empirical correlations were presented, with separate equations for the acceleration and deceleration zones.

The same technique was used by Gorshtein and Mukhlenov [86] to measure particle velocities of catalyst beads ($d = 1.5$ mm) in a conical vessel. Typical profiles obtained by these workers, who have also presented empirical correlations of their data, are shown in Fig. 4.4.

More recently Pallai [171] has reported reduced type correlations of longitudinal particle velocity profiles measured photographically in a 6 cm diameter half-sectional spouted bed. The generality of such correlations can only be determined by testing them for columns of larger diameter.

C. Radial Profile

Experimental data on radial profiles of vertical particle velocities at different levels in the spout have been reported for various solid materials and column geometries. The latter include conical [86], cylindrical [114], and conical-cylindrical [153] vessels. Typical results are shown in Fig. 4.4, where it is seen that the radial profile at any level is parabolic. These curves can be normalized according to the equation

$$v/v_0 = 1 - (r^2/r_s^2) \tag{4.23}$$

in which the maximum velocity v_0 at the axis is twice the mean velocity v_m. Confirmation of this equation is provided by Fig. 4.5, which brings together the diverse and numerous data of Gorshtein and Mukhlenov [86] for conical vessels with those of Lefroy and Davidson [114] for a flat-based cylindrical column. The data of Mikhailik and Antanishin [153] on several materials spouted in a 9.4 cm diameter column with a 60° cone also obeyed the same relationship, which thus appears to be universal.

It should be emphasized, however, that Eq. (4.23) represents time-average rather than instantaneous values of particle velocities. Volpicelli *et al.*[252] have observed that in a two-dimensional spouted bed, the particle velocities in the spout display apparently random fluctuations similar to those occurring in turbulent fluid streams. These velocity fluctuations are only to be expected, as a consequence of collisions between newly arrived particles from the annulus and the fast-moving particles in the spout.

4.2 PARTICLE MOTION IN ANNULUS AND RECIRCULATION

A. Experimental Findings

The individual particles in the annular part of the bed move vertically downward and radially inward, describing approximately parabolic paths, illustrated by the streak lines of Fig. 4.6. Axial symmetry is maintained on the average, so that, as in the spout, angular velocity components can be ignored.

Fig. 4.6. Streak photograph of air-spouted bed of wheat; tape measure reads inches. (Courtesy of National Research Council of Canada.)

1. *Interpretation of Particle Velocity at Wall Measurements*

Detailed measurements of the longitudinal (vertical) and radial (horizontal) components of particle velocity in different parts of the annulus have been reported by Thorley *et al.* [227, 142], who traced particle flow lines by direct observation against the flat wall of a 61 cm diameter semicircular column. The most important conclusion to emerge from these measurements is that in the cylindrical portion of the column, the vertical particle velocity near the wall is only slightly smaller than that near the spout at the same horizontal level. The wall velocity is thus a good approximation of the average particle velocity and hence an indicator of the total solids down flow, at a given level. In other words,

$$v_w A_a \rho_s (1 - \epsilon_a) \cong W \tag{4.24}$$

For steady state conditions, the solids down flow in the annulus at any bed level is equal to the solids up flow in the spout at that level,

$$W = v_m A_s \rho_s (1 - \epsilon_s) \tag{4.25}$$

where v_m and ϵ_s are radial mean values of particle velocity and voidage in the spout, respectively. The solids circulation rate W past any horizontal plane in the cylindrical portion of the bed may thus be determined from measurements of particle velocity v_w as observed along the wall of a round column. In the lower conical region, however, the flow lines are deflected by the sloping wall of the cone, and v_w can no longer be relied on to indicate the total solids down flow.

In the cylindrical part of the bed, v_w should decrease with decreasing bed level as a reflection of the solids cross flow dW/dz from the annulus into the spout. Thus, when small changes in spout diameter with bed level are ignored, Eq. (4.24) differentiated with respect to z yields

$$A_a \rho_s (1 - \epsilon_a) \frac{dv_w}{dz} = \frac{dW}{dz} \tag{4.26}$$

the corresponding relationship for the spout being Eq. (4.13). Dividing Eq. (4.26) by Eq. (4.24) gives

$$\frac{1}{v_w} \frac{dv_w}{dz} = \frac{1}{W} \frac{dW}{dz} \tag{4.27}$$

The right-hand side of Eq. (4.27) in fact decreases rapidly with increasing column diameter [227], and v_w itself also decreases with D_c [227]. The magnitude of the term dv_w/dz and hence its sensitivity to cross flow would therefore be expected to fall drastically with increase in D_c. This expectation is confirmed by the particle velocity at wall measurements of Thorley and

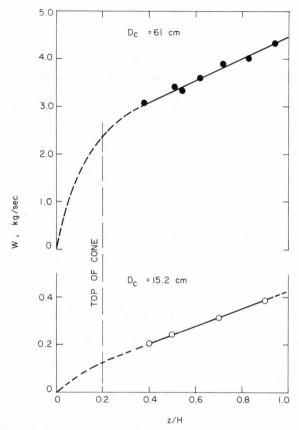

Fig. 4.7. Solids flow in annulus: Air-spouted wheat beds; $D_c/D_i = 6$, $H/D_c = 3$, $\theta = 60°$, $U_s \sim 1.1U_{ms}$. Data of Mathur and Gishler [137] for $D_c = 15.2$ cm; data of Thorley *et al.* [228] for $D_c = 61$ cm.

co-workers [227], whose 15.2 cm diameter wheat beds showed large, easily measurable vertical gradients in v_w, but whose 61 cm diameter wheat beds showed only small or nondetectable v_w gradients. Gradients in v_w are also diminished to some extent when there is a tendency of the spout to shrink and hence of the annulus to enlarge in the upper region of the bed (Fig. 5.1b, c). Such a tendency was observed by Thorley *et al.* in their larger column. To get a precise answer for the cross flow dW/dz, it is then necessary to include A_a within the differential, which thus becomes $d(A_a v_w)$, in Eq. (4.26).

2. *Down Flow and Cross Flow of Solids*

Typical solids flow curves, derived from v_w data of Mathur and co-workers using Eq. (4.24), are shown in Fig. 4.7. The linearity of W with z

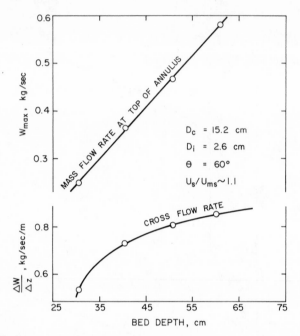

Fig. 4.8a. Effect of bed depth on solids circulation calculated from data of Thorley *et al.* for wheat [227, 228].

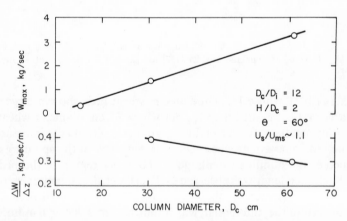

Fig. 4.8c. Effect of column diameter on solids circulation calculated from data of Thorley *et al.* for wheat [227, 228].

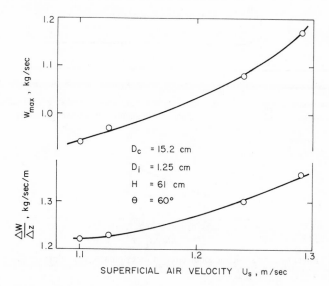

Fig. 4.8b. Effect of gas flow rate on solids circulation calculated from data of Thorley *et al.* for wheat [227, 228].

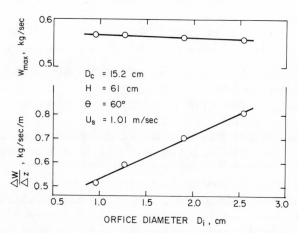

Fig. 4.8d. Effect of orifice diameter on solids circulation calculated from data of Thorley *et al.* for wheat [227, 228].

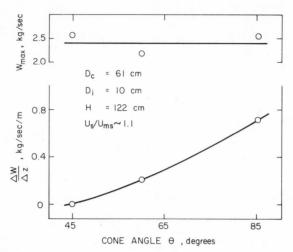

Fig. 4.8e. Effect of cone angle on solids circulation calculated from data of Thorley *et al.* for wheat [227, 228].

above the cone is characteristic of spouted beds, so that the cross flow rate per unit height in the cylindrical region is constant and can simply be expressed as $\Delta W/\Delta z$ for this region. The sharp change of slope which occurs in the cone, especially for the larger column, is symptomatic of the continually collapsing spout–annulus interface near the gas inlet. The total solids flow or maximum solids circulation rate W_{max} is simply the value of W at the top of the bed.

A selection from the extensive experimental results of Thorley *et al.* is plotted in Fig. 4.8. The following trends are demonstrated by these plots:

(a) Increasing the bed depth increased both the total solids flow rate and the solids cross flow rate.

(b) For a given bed depth, increasing the spouting gas rate increased both the total solids flow and the cross flow.

(c) For geometrically similar beds, increasing the column diameter increased the total solids flow but had little effect on the cross flow rate above the cone.

(d) At a given gas flow, increasing the orifice size increased the cross flow rate in the cylindrical region without significantly affecting the total solids flow.

(e) Increasing the cone angle from 45° to 85° increased the cross flow rate in the upper region of the bed for the 61 cm diameter column. The same effect was, however, not observed in the 15.2 cm diameter column, where cone angle had little influence on cross flow. In neither column did cone angle have any significant effect on total solids flow.

The latter result is somewhat at odds with the results of Elperin *et al.* [55], who reported that a cone angle of 40°–45° was the optimum for maximizing solids circulation rate. However, it should be noted that the beds studied by these workers remained wholly or mainly in the conical section of the spouting vessel.

Attempts to generalize on experimental results for particle velocity at the wall have been made by several investigators, and these are summarized in Table 4.1. The correlations are useful for predicting changes in solids circulation rates with changing operating conditions, but cannot be relied upon outside the range of parameters for which each was developed. Becker's [15] equation can also be used to estimate $\Delta v_w/\Delta z$ and hence the cross flow.

3. Cycle Time

The distribution of times taken by particles to make the journey from the top of the annulus downward and back again to their starting points is of considerable interest in processes where solids mixing is important, as well as in distinctively cyclical operations such as granulation (Chapter 11). Since the proportion of time spent by a particle in the spout is insignificant compared with that spent in the annulus, particle cycle times can be deduced from solids flow patterns in the annulus.

The distribution of cycle times was determined by Thorley *et al.* [228] from the particle flow lines measured in their 61 cm column, mentioned earlier. This is plotted in Fig. 4.9, which represents the time taken by different cumulative weight fractions of solids to complete one cycle. Although some short circuiting of particles occurred at the top of the bed, over 90% of the wheat took 60 sec or longer to complete one cycle. From Fig. 4.7, the solids circulation rate at the top of the same bed, W_{max}, which is a measure of the bed turnover rate, was 2.2 kg/sec, while from the geometry and bulk density of the bed, the solids holdup was 223 kg. The average cycle time, assuming no dead solids, was therefore 223/2.2 or 101 sec, which is in good agreement with the value of 99 sec obtained by measuring the area under the curve of Fig. 4.9.

Becker measured the maximum cycle time by dropping a marked pellet onto the bed adjacent to the wall and noting the lapse of time to its reappearance at the top of the bed. His data for wheat beds with $D_c = 15$–61 cm, $H = 0.6$–1.8 m, $D_i/D_c = 0.17$, and cone angles of 90° and 180° were correlated by the equation

$$t_c = 21(H/g)^{1/2}(H_m/D_c) \tag{4.28}$$

This equation must be used in conjunction with a relationship for H_m, such as those discussed in Chapter 6. For example, by Eq. (6.5), $H_m/D_c \sim 8$ for

TABLE 4.1 Empirical Correlations for Particle Velocity at Wall[a]

Investigators	Correlation	Range of supporting data
Thorley et al. [227]	$W_{avg} = v_w A_a(1 - \epsilon_a) = K(D_i/D_c)^{-0.25}[u_i/(u_i)_{ms}]^{1.23}(H/D_c)^{1.0}$ W_{avg} in kg/sec with $K = 0.563$ for 61 cm diameter column and $K = 0.068$ for 15.2 cm diameter column; 60° cone in each case. v_w is average particle velocity at wall above cone.	Wheat $D_c = 15.2$ cm and 61 cm $D_i = 1.25$–10.2 cm $\theta = 60°$ $H = 30.5$–183 cm
Becker [15]	$(v_w{}^2/gH)^{0.4}(U_a/U_m)^2 = B(z/H_m)^b$ For $z/H_m < 0.25$, $B = 0.055$ and $b = 1$; for $z/H_m > 0.25$, $B = 0.21$ and $b = 2$. U_a/U_m can be estimated from Eq. (2.17) with $U_m \sim U_{mf}$ and $n = 1.65$ for wheat, and H_m by Eq. (2.44). Measurements of v_w were made at different bed levels z at the minimum spouting condition.	Wheat 15.2 cm diameter column with flat base $D_i/D_c = 0.065$–0.285
Shigeo [213]	$v_w/U_{ms} = 7.6 \times 10^{-3}(gD_c/U_{ms}^2)^{-0.4}(D_i/D_c)^{-0.7}(U_s/U_{ms})^{1.7}$ v_w is average particle velocity over the height of the annulus.	Rice ($d_p = 3.8$ mm) $D_c = 18$ cm $H = 18$–72 cm $D_i = 0.9$–4 cm U_s up to 1.5U_{ms}

Matsen [145a]	$(v_w)/(v_w)_{ms} = U_s/U_{ms}$	Billings green coke ($-8 + 14$ mesh) $D_c = 14$ cm $D_i = 1.3$ cm U_s up to $1.6 U_{ms}$
	v_w measured near top of annulus.	
Abdelrazek [1]	$\log(v_w/U_{ms}) = 3.42 \times 10^{-4}(\mathrm{Re})(H/D_c) - 1.543$	Wheat (data of Thorley *et al.* [227]) $D_c = 15.2$ and 61 cm $D_i = 1.25-10.2$ cm $\theta = 60°$ $H = 30.5-183$ cm
	where $\mathrm{Re} = d_p \rho_g U_s/\mu$	
	v_w is particle velocity at wall at top of annulus.	
Németh and Pallai [166]	$(v_w) = (v_w)_{Hm}(H/H_m)^{1/3}$	Synthetic resin, activated carbon, glass beads ($d_p = 0.6-2.5$ mm) $D_c = 6.0$ cm $D_i = 2-10$ mm $H/H_m^1 = 0.31-0.78$
	and	
	$[v_w]_{(u_i)_1} = [v_w]_{(u_i)_2}\left[\dfrac{(u_i)_1}{(u_i)_2}\right]^{3/2}$	Synthetic resin, activated carbon ($d_p = 0.8-2.0$ mm) $D_c = 6.0$ cm $D_i = 2-10$ mm $(u_i)_1/(u_i)_2$ up to 1.6
	v_w is average particle velocity over the height of the annulus.	

[a] Spouting fluid, air.

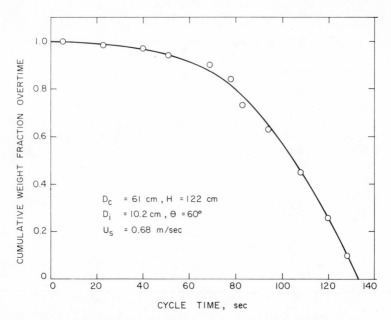

Fig. 4.9. Cycle time distribution for solids in air-spouted wheat [228]. Cumulative solids fraction is measured radially from wall inward. Reproduced by permission of *The Canadian Journal of Chemical Engineering* **37**, 184 (1959), Fig. 6.

the 1.22 m deep bed depicted in Fig. 4.9, and therefore by Eq. (4.28), $t_c \sim 60$ sec. This answer falls far short not only of 133 sec, the maximum cycle time in Fig. 4.9, but even of the average cycle time of 100 sec, thus demonstrating the hazards of using an empirical relationship such as Eq. (4.28) outside the range of parameters (in this case, cone angle) for which it has been derived. Figure 4.8 suggests that for Becker's cone angles of 90° and 180° there was considerably more cross flow in the upper region of the bed, and hence more particle short circuiting, than for the 60° cone angle of Fig. 4.9.

Even more limited cycle time correlations have been reported by Gay *et al.* [68] for spouting of whole peanuts. These investigators measured the maximum and two intermediate cycle times, the smaller obtained by introducing a tracer particle at the top of the bed halfway between the wall and the spout, and the larger by averaging ten successive random cycle times. Their three relationships predict dependencies of cycle time on bed parameters which are very different from each other as well as from Eq. (4.28).

B. Theoretical Predictions

The mass and momentum balance method for the spout described earlier yields both W [e.g., via Eq. (4.25)] and dW/dz as a function of z. Lefroy and Davidson [114], who first proposed this method, computed W at various bed levels for conditions corresponding to their experimental runs in flat-based columns. They found their computed values of W to vary roughly linearly with z, in agreement with Fig. 4.7 above the conical regions. For the beds corresponding to Figs. 4.2 and 4.3, the same qualitative trend with respect to z was noted by us in values of W computed by the mass and momentum balance method. The absolute values of W computed by this method have, however, been found to be correct only in order of magnitude and, as already noted for both v and u_s, very sensitive to the assumed spout contour.

An entirely independent model for predicting cross flow was developed, again by Lefroy and Davidson [114], from an analysis of the mechanism of particle to particle collisions at the spout–annulus interface or "spout wall." The starting point of their analysis is the observation originally made by Thorley *et al.* [228] from their high speed motion pictures of the spout, that near the base of the bed, the spout wall was continually collapsing and particles were being swept into the high velocity jet, while further up the bed, lateral transfer of particles across the spout boundary appeared to occur as a result of collisions between the upward moving particles in the spout against the particles forming the spout wall. The model concerns the upper part of the bed only. The radial velocity v_r of a particle leaving the annulus is by this analysis related to the vertical velocity v_a of a rising spout particle. The glancing collision of the latter particle with the spout wall is considered to result in momentarily entraining the annulus particle into the spout. The derived relationship is

$$v_r/v_a = \pi^2 e(1 + e)/64 \tag{4.29}$$

where e is the coefficient of restitution of the particles. The center of particles having velocity v_a is assumed on the average to be one particle diameter d_p from the spout wall. Therefore, from Eq. (4.23), assuming $d_p \ll D_s$, it follows that

$$v_a/v_m = 2v_a/v_0 = 8d_p/D_s \tag{4.30}$$

Eliminating v_a between Eqs. (4.29) and (4.30) gives

$$v_r/v_m = \pi^2 e(1 + e)d_p/8D_s \tag{4.31}$$

The momentary flux of solids into the spout at a given bed level, assuming that the spout voidage ϵ_s at that level is invariant across the spout, is $v_r \rho_s(1 - \epsilon_s)$. However, only a fraction of these solids will be permanently entrained, the remainder rebounding back into the annulus as a result of impact with particles already in the spout. The fraction permanently entrained depends on the difference in solids concentration between the annulus and the spout, and is given by

$$\frac{(1 - \epsilon_a) - (1 - \epsilon_s)}{1 - \epsilon_s} \equiv \frac{\epsilon_s - \epsilon_a}{1 - \epsilon_a}$$

Since the spout wall area for a differential height dz of bed is $\pi D_s \, dz$, it follows that the permanent entrainment rate or net cross flow rate of particles into the spout per unit height of bed is

$$\frac{dW}{dz} = v_m\left(\frac{v_r}{v_m}\right)\rho_s(1 - \epsilon_s)\left(\frac{\epsilon_s - \epsilon_a}{1 - \epsilon_a}\right)\pi D_s \tag{4.32}$$

Substituting Eq. (4.31) into Eq. (4.32) yields

$$\frac{dW}{dz} = \frac{\pi^3 e(1 + e)d_p v_m \rho_s(1 - \epsilon_s)(\epsilon_s - \epsilon_a)}{8(1 - \epsilon_a)} \tag{4.33}$$

Lefroy and Davidson were not able to test Eq. (4.33) directly since they were unable to obtain reliable measurements of ϵ_s in their experiments (see Chapter 5). Instead, they showed that values of the ratio v_r/v_m from Eq. (4.31) were similar to those obtained from Eq. (4.32), using in the latter equation the values of v_m, ϵ_s, and dW/dz determined by the method of mass and momentum balances in the spout. We have, however, evaluated dW/dz at various levels in the upper half of the bed directly from Eq. (4.33) for the conditions of both Figs. 4.2 and 4.3, estimating the experimental values of v_m from these figures and the corresponding values of ϵ_s at matching bed levels from Fig. 5.3. In both cases, the values of dW/dz thus calculated were typically almost an order of magnitude greater than the experimental values of $\Delta W/\Delta z$ in the upper regions of the beds, as given by Fig. 4.7. While it is true that Eq. (4.33) is very sensitive to changes in v_m and ϵ_s, which must therefore be known more accurately than is generally the case to evaluate dW/dz, the collision model is still only a rough approximation of reality which, in addition to its explicit simplifications, neglects such factors as friction between particles and roughness of the spout–annulus interface. Its principal virtue is that it does not require a knowledge of the spout contour.

The only theoretical attempt at explaining the concentrated cross flow behavior at the bottom of the bed has been made by Volpicelli et al. [252]. These workers have shown that the very high lateral particle velocities which occur only a few centimeters above the gas inlet are predictable from the principles of plasticity applied to gravity flow of bulk solids in a converging channel, are accentuated by conical as opposed to flat bases, are independent of spouting velocity for a given bed, and are attenuated by increasing the column diameter up to about 15 cm, becoming independent of D_c thereafter. The constant value of the maximum lateral velocity for conical-cylindrical beds with $D_c \geq 15$ cm was computed from the experimental data of other workers [137, 15, 127] to be about 7 cm/sec, in agreement with the value obtained in their own experiments with two-dimensional spouted beds.

4.3 GROSS MIXING BEHAVIOR

A. Experimental Findings

Information concerning gross mixing behavior of solids comes mainly from stimulus–response experiments where the residence time distribution of particles fed into and discharged from the bed at a steady rate is measured. The range of conditions over which such experiments have been carried out, by three different groups of investigators [16, 104, 9], are shown in Table 4.2. Typical data are plotted in the form of an "F curve" in Fig. 4.10, while all the available results are shown in Fig. 4.11 as a plot of log $I(\theta)$

TABLE 4.2 Experimental Conditions Used for Obtaining
Stimulus–Response Data in Fig. 4.11

	Becker and Sallans [16]	Kugo *et al.* [104]	Barton *et al.* [9]
Material	Wheat	Wheat	Coke, $-6 +10$ mesh
D_c (cm)	23	15	14.6
D_i (cm)	4	0.6–1.7	~ 2.5
H (cm)	91	8–12	~ 40
θ (deg)	120	82	90
$\bar{t}$ (min)	12–47	25–40	~ 10–40
Technique	Pulse of dyed wheat injected at bed top, concentration of colored particles in bed measured periodically.	Feed changed to colored particles, their concentration measured in wheat discharging from opening in the conical base.	Stepwise change imposed on solids feed as by Kugo *et al.*, but solids discharged through an overflow pipe at bed top diametrically opposite feed inlet point.

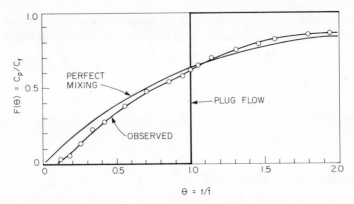

Fig. 4.10. Typical output response to step change in solids input [104]. System: air–wheat; $D_c = 15$ cm, cone angle $= 82°$.

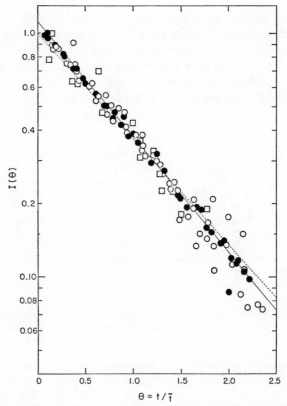

Fig. 4.11. Correlation of solids mixing data [60]. Data of: ● Kugo *et al.* [104]; □ Becker and Sallans [16]; ○ Barton *et al.* [9]; ---, perfect mixing [Eq. (4.34)]; ——, Eq. (4.35).

versus $\theta = t/\bar{t}$. It is seen that the data in Fig. 4.11 rectify fairly well according to the relationship.

$$I(\theta) = C_b/C_{b0} = e^{-\theta} \tag{4.34}$$

which corresponds to perfect mixing.

It should, however, be pointed out that Barton *et al.* [9], who fitted their own experimental data by linear regression, obtained the following empirical equation within $\pm 10\%$ mean deviation:

$$I(\theta) = \exp[-(1/0.92)(\theta - 0.10)] \tag{4.35}$$

which is represented by the solid line in Fig. 4.11. By invoking the mixed-flow models of Cholette and Cloutier [39] and of Levenspiel [119], the interpretation of Eq. (4.35) ultimately arrived at by Quinlan and Ratcliffe [186] is that 8–10% of the total spouted bed volume is in plug flow, while the remainder is perfectly mixed, deadwater and bypassing being negligible. The conclusion that solids mixing in spouted beds is nearly but not quite perfect was also arrived at independently by Kugo *et al.* [104] and most recently by Pallai and Németh [174, 175]. All the same, it is apparent from Fig. 4.11 that the perfect mixing assumption would be a good approximation for most practical purposes. This conclusion is, of course, subject to the mean residence time exceeding some minimum value, the feed and discharge points being so located as to preclude any obvious short circuiting [28], and the cone angle being sufficiently steep to effect smooth flow of solids at the base.

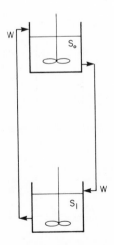

Fig. 4.12. Twin CST model used by Chatterjee [38].

Two other approaches to measuring the gross mixing behavior of solids have been reported, both concerned with the intermixing of solids within the bed itself rather than with mixing between the bed-solids and fresh feed in a continuous system. Chatterjee [38] filled the upper one-third of his 15 cm diameter × 25 cm deep spouted bed with colored particles and measured the change in tracer concentration with time at the top of the bed. A similar procedure has been used for fluidized beds [263, p. 290]. He interpreted his results in terms of the simple twin completely stirred tank (CST) model depicted in Fig. 4.12. The integrated mass balance equation for this case is

$$\ln \frac{1 - x_e}{x_0 - x_e} = W\left(\frac{1}{S_0} + \frac{1}{S_1}\right)t \tag{4.36}$$

where S_0 and S_1 are the weights of the initially traced and untraced bed regions, respectively; x_0 the tracer weight fraction in the upper part of the bed; and x_e the equilibrium tracer concentration ($= \frac{1}{3}$ in Chatterjee's experiments). Semilogarithmic plots of $(1 - x_e)/(x_0 - x_e)$ versus $[(1/S_0) + (1/S_1)]t$ were made, and the slopes evaluated as W, the solids circulation rate.

The values of W thus obtained for beds of sand, mustard seed, sago, and coal ($d_p = 1\text{–}3$ mm, $\rho_s = 1\text{–}3$ Mg/m³), with $\theta = 60°$, $D_i = 9.52$ mm, and $U_s \sim (1.5\text{–}2)U_{ms}$, ranged between 0.17 and 0.53 kg/sec. These values are similar in magnitude to the solids circulation rates for wheat determined

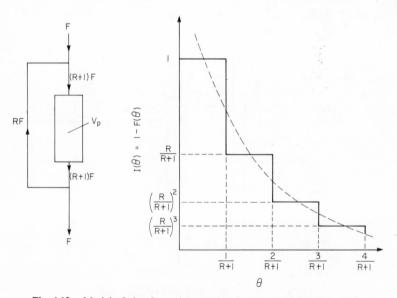

Fig. 4.13. Model of plug flow with recycle, after Levenspiel [118, p. 291].

from the measurements of particle velocity at the wall in a 15.2 cm diameter column [137, 228] reported in the previous section. Chatterjee correlated his data by a dimensional equation which is equivalent to

$$W = 0.24(U_s/U_{ms})(d_p^{0.2}/\rho_s^{0.17}) \tag{4.37}$$

with W in kilograms per second, d_p in millimeters, and ρ_s in megagrams per cubic meter. Equation (4.37) shows that while particle size and density have a small effect on solids circulation rate, the dominating variable for a given column diameter and bed depth is the gas flow rate. A similarly strong dependence of W on U_s has already been indicated by Fig. 4.8b.

Qualitatively, the above findings by Chatterjee appear to be quite plausible, but the quantitative reliability of his circulation rate results has been put into doubt by the criticism of Mann and Crosby [132]. These workers point out that the mixing data for coal beds reported in Chatterjee's paper correspond to concentrations x_0 in the range 0.3439–0.3340, i.e., within only 3% of the equilibrium value. The results of W based on such a limited range of concentrations taken so close to the equilibrium condition would be subject to large errors. It appears that the measurements of x_0 in Chatterjee's experiments were delayed too long after the introduction of the tracer. Mann and Crosby have also criticized Chatterjee's simple model as being unrealistic and have proposed more sophisticated models, which are discussed later.

The second approach mentioned involves a statistical study of solids circulation, carried out by Galkin et al. [67], in a two-dimensional 40°-cone-based spouted bed of silica gel particles ($d_p = 2.5$ mm and 4.3 mm). Two tracer particles opaque to X rays were introduced and motion-picture sequences taken. A detailed analysis of the results was made, and turbulence type parameters such as correlation times and mixing lengths were computed for the particle motion. Effective particle diffusion coefficients were also evaluated and these were found to increase with spouting velocity, finally leveling off at about $U_s \sim 2U_{ms}$, where the bed was described as attaining a state of maximum statistical uncertainty. A similar trend had been previously observed by these workers for cylindrical fluidized beds, the effective particle diffusivities of which were of the same order of magnitude as those for spouted beds.

B. Modeling

1. Minimum Residence Time Requirement

Consider the simple recycle model illustrated in Fig. 4.13. The plug flow volume V_p is assumed to represent the annulus and hence the bulk of the bed, while the recycle action is assumed to occur in the spout, the volume of which is taken as negligible relative to that of the annulus. Solids short circuiting from the annulus to the spout is ignored, and hence the model is conservative with respect to the actual mixing behavior of a spouted bed. The theoretical output response to a step input change is incremental, rather than smooth, as shown in Fig. 4.13, but the actual output from a real system in which some axial dispersion and cross mixing occurs is more apt to be smoothed out, as illustrated by the dotted line in the same figure. This dotted line is bracketed by the locus of the points below it, given by

$$I(\theta, R) = \left(\frac{R}{R+1}\right)^{(R+1)\theta} \tag{4.38}$$

and by the locus of the points above it, given by

$$I(\theta, R) = \left(\frac{R}{R+1}\right)^{(R+1)\theta - (1/R+1)} \tag{4.39}$$

When plotted as $\ln I(\theta)$ versus θ, Eqs. (4.38) and (4.39) yield parallel straight lines with a slope of $(R+1)\ln[R/(R+1)]$, the value of the slope being indicative of the degree of mixing [146]. A steep slope, which corresponds to small values of R, represents conditions approaching plug flow, while for a slope of $-1(R \to \infty)$, Eqs. (4.38) and (4.39) both reduce to Eq. (4.34) for perfect mixing. The plot in Fig. 4.14, however, shows that even for R as

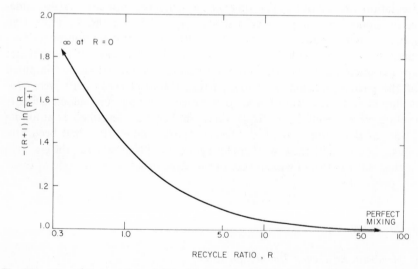

Fig. 4.14. Approach to perfect mixing with increasing recycle ratio according to Eqs. (4.38) and (4.39).

low as 5, the mixing behavior according to the above criterion becomes quite close to perfect mixing—at least close enough for most practical purposes. For example, the empirical equation [Eq. (4.35)] as plotted on Fig. 4.11 has a slope of −1.09, which corresponds to a value of $R = 5.0$ on Fig. 4.14. A value of $R = 5$ therefore seems suitable as the basis for obtaining an order of magnitude estimate of the minimum residence time necessary to achieve "near-perfect" mixing.

The cycle time in the recycle model is given by

$$t_c = \frac{V_p}{(R + 1)F} \tag{4.40}$$

while the mean residence time is given by

$$\bar{t} = V_p/F \tag{4.41}$$

Combining Eqs. (4.40) and (4.41) yields

$$\bar{t} = t_c(R + 1) \tag{4.42}$$

For $R = 5$, $\bar{t} = 6t_c$, which is thus the desired minimum residence time.

The cycle time t_c has been discussed earlier in this chapter (Section 4.2). From experimental measurements it can be obtained as

$$t_c = W_b/W_{max} \tag{4.43}$$

Table 4.3 shows typical values of t_c from wheat spouted experiments, and

TABLE 4.3 Predicted Values of Minimum Residence Time
Required to Achieve Good Mixing[a]

Ref	D_c (cm)	θ (deg)	D_c/D_i	H/D_c	W_b (kg)	W_{max} (kg/sec)	$t_c = W_b/W_{max}$ (sec)	$\bar{t} = 6t_c$ (sec)
[228]	15.2	60	6	2	3.2	0.33	10	60
[228]	30.5	60	12	2	26.7	1.36	20	120
[227]	61.0	60	6	2	223	2.2	101	606

[a] Based on cycle time data for wheat beds; $U_s \sim 1.1 U_{ms}$.

the corresponding values of $\bar{t}$ calculated from Eq. (4.42), assuming $R = 5$. It is seen that in the largest column considered ($D_c = 61$ cm), the estimated minimum residence time requirement is about 10 min, while for the smaller columns, the computed values of $\bar{t}$ are considerably lower. The observed near-perfect mixing behavior in columns up to 23 cm diameter for $\bar{t} > 10$ min (Fig. 4.11 and Table 4.2) is therefore consistent with the above predictions.

Referring to Eq. (4.43), it is expected that increasing the column diameter

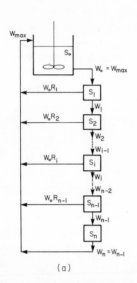
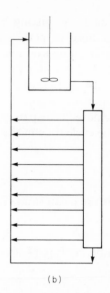

(a)   (b)

Fig. 4.15. Solids circulation models for spouted beds, after Mann and Crosby [132]. (a) CSTs-in-series model with short circuiting to mixing zone; (b) plug flow model with short circuiting to mixing zone. Reprinted with permission from *Ind. Eng. Chem. Process Des. Develop.* **11,** 314 (1972). Copyright by the American Chemical Society.

beyond 61 cm will sharply raise the minimum value of $\bar{t}$ above 10 min, since for a fixed value of H/D_c, the bed volume (and hence W_b) is proportional to D_c^3, while according to Fig. 4.8c, W_{max} increases only linearly with D_c. On the other hand, increasing the bed depth for a fixed column diameter should have only a small effect on the residence time requirement, since it is seen from Fig. 4.8a that W_{max} is roughly proportional to H and hence also to W_b, so that from Eq. (4.43), t_c would remain substantially unchanged.

2. Solids Circulation

While the recycle model of Fig. 4.13 does serve the purpose of providing a practical guideline on residence time requirement, it oversimplifies the actual solids circulation in a spouted bed and is, therefore, of limited utility. A more realistic model, which takes into account solids cross flow from annulus to spout as well as holdup in the spout, has been proposed by Mann and Crosby [132], whose analysis, however, concerns the mixing behavior of a batch rather than a continuous system. Their proposed model is shown in Fig. 4.15a. The spout plus fountain are considered a completely stirred tank (CST) containing S_0 kg of solids at any instant, while the annulus is considered as n completely stirred tanks (CSTs) in series, with short circuiting to the spout from each CST above the nth (i.e., the bottom CST). For a batch operation in which part of the solids are initially tracer particles, total solids and tracer solids balances were made on each CST. Numerical solution of the resulting equation, for assigned values of n, S_0/W_b, and ΣR_i, yields the tracer concentration x_0 at the top and x_n at the bottom of the bed, as a function of time. Figure 4.16 shows illustrative results for different values of n, with S_0/W_b taken as 0.15 (although a value in the range of 0.02–0.06 would be more realistic; see Fig. 5.3) and ΣR_i as $\frac{1}{3}$, and assuming that (a) the n CSTs representing the annulus contain equal weights of solids, (b) short circuiting is uniformly distributed among $n - 1$ CSTs, and (c) the top one-third of the annulus initially contains only tracer particles, while the bottom two-thirds is devoid of tracer. It is seen that both x_0 and x_n reach their equilibrium value of $\frac{1}{3}$ very rapidly and tend to cycle around this value thereafter, the tendency being more pronounced for larger values of n. These results indicate the importance of sampling at different locations in the spouted bed in tracer experiments designed to evaluate the proposed model, as well as the necessity for taking samples as soon as possible after tracer injection.

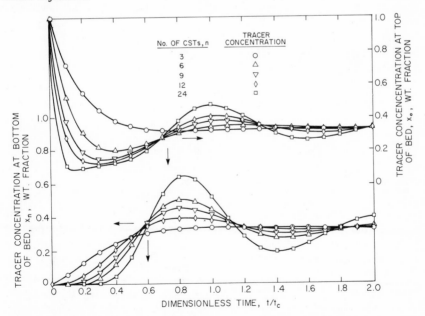

Fig. 4.16. Evaluation of the model of Fig. 4.15a [132]. Reprinted with permission from *Ind. Eng. Chem. Process Des. Develop.* **11**, 314 (1972). Copyright by the American Chemical Society.

While the concept of representing the annulus by a series of CSTs leads to a generalized model, the movement of annular solids in real spouted beds is undoubtedly close to plug flow. The simpler model of Fig. 4.15b, which corresponds to $n \to \infty$ in the model of Fig. 4.15a, would therefore seem to be adequate. It should also be noted that the generalized model of Fig. 4.15a collapses to the model used by Chatterjee (Fig. 4.12), though with the unrealistic conditions that n equals unity, short circuiting is absent, and solids holdup in the spout region is insignificant [132].

Methods for determining the cycle time distribution of solids within closed continuously circulating systems, such as a spouted bed, have recently been developed further by Mann and Crosby [133].

5 | *Bed Structure*

5.1 SPOUT SHAPE

The fact that in a steadily spouting bed the spout assumes a stable shape implies the existence of a state of dynamic equilibrium between the various forces acting on the spout–annulus interface. Since these forces arise from the movement of both solids and gas, the shape of the spout can provide some valuable clues to an understanding of the entire dynamics of spouted beds. Spout shape has therefore received considerable attention.

A. Experimental Observations

Spout shape observations have usually been made against the flat transparent face of sectional columns, either "semicircular" or "two-dimensional." A variety of shapes have been observed under different experimental conditions and these are illustrated in Fig. 5.1, which shows notional sketches of the different types of spout shapes reported in the literature. The experi-

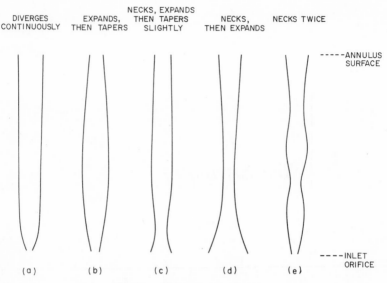

Fig. 5.1. Observed spout shapes.

mental conditions associated with each type are listed in Table 5.1. It should be noted that the last two shapes in Fig. 5.1 have been observed only in two-dimensional columns and may not be representative of spout shape in a normal bed. Doubt concerning distortion of the spout in a half-sectional column by the flat face has been dispelled to a large extent by Mikhailik [152], who made parallel measurements in half and full cylindrical columns, using for the latter the piezoelectric technique mentioned in Chapter 4. From his results of spout diameter and its variation with bed level, Mikhailik concluded that spout shapes obtained in the two columns, for a variety of solid materials, were quite similar.

Examination of Fig. 5.1 and Table 5.1 shows that the pattern of change of diameter with bed level is obviously not related to the inlet orifice diameter, nor indeed to any other variable, in any simple manner. Particularly striking is the contrast between shape (a) in Fig. 5.1, where the spout shows a marked expansion in the region immediately above the entrance orifice, and shape (c), where the spout suddenly contracts over this region.

A substantial change in spout diameter in the vicinity of the gas inlet is a matter of considerable importance since it directly affects the longitudinal profile of gas velocity in the spout, and consequently also influences particle velocity and voidage profiles. This particular feature of spout shape has, however, not received much attention, and the inability to make predictions in this respect has proved to be a major difficulty in the development of

TABLE 5.1 Experimental Conditions for Fig. 5.1

Spout type	Column type	D_c (cm)	D_i (cm)	θ (deg)	H (cm)	Solids used	Comments	Ref.
(a)	Semicircular	15.2	1.25	85	63	Wheat	Spout diverges at angle of $\simeq 18°$ (see Fig. 1.2a).	[137]
	Conical	—	2.0–4.2	20–50	10–50	Fertilizer fractions	Spout diverges at angles between 10° and 19°.	[231]
	Conical	—	—	20–70	—	—	Angle of spout divergence remains at 5°–7° over a wide range of conditions.	[159]
	Semicircular	15.2	0.9–1.9	85	30–61	Wheat		[228]
	Semicircular	15.2	1.27	60	18–25	Wheat	Spout diverges quickly to nearly constant diameter.	[127]
	Circular	9.4	1.5	45	<20	Wheat, millet, polystyrene, silica gel		[152]
(b)	Semicircular	15.2	1.27	60, 90, 180	18–25	Polystyrene	Note that with a larger orifice, spout shape changed to type (c).	[94]
	Semicircular	61	5.1–7.6	60	91–183	Wheat	—	[228]
	Circular	9.4	1.5	45	>20	Wheat, millet, polystyrene, silica gel	Tapering of spout diameter near top occurred only for $H > 20$ cm.	[152]
(c)	Semicircular	—	2.55–5.1	180	75–180	Kale seed, polyethylene, peas	Spout walls near base were continually collapsing. With wheat, maximum spout diameter attained at $z \simeq 0.4H$.	[114][a]
	Semicircular	61	10.2	60	183	Wheat		[228]
(d)	Two-dimensional with conical face	—	$\simeq 6$	70	$\simeq 20$	"Beads" of 1.0–1.5 mm diameter	See Fig. 1.2b. Angle of the upper diverging part of the spout varied between 8.5° and 18°, depending upon air flow rate.	[78]
(e)	Two-dimensional with rectangular face	1.6×20	0.4×1.6	180	20–30	Ceramic chips, glass beads	Disturbances developed at the bottom of the spout and moved upward like ripples.	[252]

[a] In these experiments upper spout diameter exceeded D_i, as in spout type (a).

theoretical models concerned not only with particle motion in the spout, as pointed out in Chapter 4, but also with spout voidage.

The findings of a recent visual study of spout shape in the gas entrance region carried out at UBC [259] are therefore worth recording despite the fact that only one column—15 cm diameter half-sectional with a 60° conical base—was used. Spout diameters at various vertical positions were measured in beds of copper slag, Ottawa sand, polyglycol cubes, and polyethylene cylinders (mean $d_p = 0.71$–2.82 mm, $\rho_s = 1.02$–2.89 Mg/m^3). The other variables studied were bed depth (6.3–73.5 cm), air inlet diameter (1.27–3.81 cm), and air flow rate (up to $2U_{ms}$). The spout shape in beds of fine Ottawa sand (mean $d_p = 0.71$ mm) was found to be relatively insensitive to changes in these variables, but the results for the other three materials showed that:

(1) Necking of the spout occurred in the entrance region only in shallow beds (< 35 cm deep), and at low air flow rates. Expansion generally occurred in deeper beds, or in shallow beds at high air flow rates.

(2) When the spout necked, the neck diameter was roughly equal to 0.86 times the orifice diameter for all three solid materials, regardless of bed depth and air flow rate. Also, the vertical distance from the air entrance at which the spout diameter reached its minimum value was found to be inversely proportional to orifice diameter, the approximate value of the proportionality constant being 1 cm^2.

(3) When the spout expanded, the increase in spout diameter with bed level occurred abruptly in some cases and gradually in others, and no simple pattern with respect to the independent variables could be discerned.

For beds contained in wholly conical vessels, spout shape observations made with a piezoelectric probe, using different cone angles ranging between 20° and 70°, led Mukhlenov and Gorshtein [159, 160] to the conclusion that the spout always expands along the bed height [shape (a) in Fig. 5.1] with a divergence angle of 5°–7°. The range of experimental conditions, namely, solid properties, bed depths, orifice size, and air flow rates, on which the above conclusion is based has not been given by these workers, but their explanation of the simple observed shape implies that the above variables do not play a role. Mukhlenov and Gorshtein suggest that the shape of the spout is determined in accordance with the principle of least resistance, and refer to the work of Gibson [69], who showed that a conical diffuser with an angle of 5° 35′ offers minimum resistance to fluid flow. It is difficult to reconcile this explanation with the variety of spout shapes observed in cylindrical spouted beds. The analogy with a diffuser is in any case an oversimplification, since it disregards the important effect on spout shape

TABLE 5.2 Empirical Correlations for Mean Spout Diameter

Source	Correlation		Range of supporting data
Malek et al. [127]	$D_s = (0.115 \log_{10} D_c - 0.031)G^{1/2}$ where D_s and D_c are in inches and G in lb/hr-ft^2.	(5.1)	$D_c = 10$–61 cm $D_i = 0.6$–10 cm $H = 15$–183 cm $G = 0.35$–1.8 kg/sec m^2 Solids: wheat, barley, millet, rice, corn, flax seed, polyethylene, polystyrene $d_p = 1.3$–7.3 mm, $\rho_s = 0.92$–1.4 Mg/m^3 Average deviation from equation, $\pm 6\%$.
Mikhailik [152]	$D_s = 10(0.115 \log_{10} D_c - 0.031)(G/\rho_s)^{1/2}$ Same units as (5.1), ρ_s in pounds per cubic foot, or $D_s = 14.5(0.115 \log_{10} D_c - 0.192)(G/\rho_s)^{1/2}$ with D_s and D_c expressed in millimeters, G in kg/hr-m^2, and ρ_s in Mg/m^3.	(5.2)	$D_c = 9.4$ and 14 cm $D_i = 1.5$ and 2.5 cm $H = 5$–25 cm $G = 0.71$–5.25 kg/sec m^2 Solids: wheat, millet, polystyrene, silica gel, slag beads, pig iron pellets $d_p = 1.5$–4 mm, $\rho_s = 1.4$–7.8 Mg/m^3 The data of Malek et al. [127] are also claimed to be correlated by this equation [152].
Abdelrazek [1]	$D_s = 0.315 D_c [U_s/(gH)^{1/2}]^{0.33}$ Any consistent units	(5.3a)	$D_c = 5$–10 cm $D_i = 0.42$–0.85 cm $H = 7.6$–33 cm $G = 0.13$–0.43 kg/sec-m^2 Solids: glass beads only $d_p = 0.52$ and 0.61 mm, $\rho_s = 2.46$ Mg/m^3
	$D_s = 0.346 D_c [U_s/(gH)^{1/2}]^{0.50}$ Any consistent units	(5.3b)	Same as for Eq. (5.1) Maximum error, 22%.
McNab [148]	$D_s = \dfrac{0.037 G^{0.49} D_c^{0.68}}{\rho_b^{0.41}}$ where D_s and D_c are in feet, G in lb/hr-ft^2, and ρ_b in pounds per cubic foot.	(5.4)	Same as for Eq. (5.1) Average deviation from equation, 5.6%; maximum deviation, 17.8%.

of solids movement which obviously does come into play, at least under certain conditions.

B. Empirical Correlations

In cylindrical beds with a short conical base, the major adjustment in the spout diameter most commonly occurs in the region immediately above the inlet orifice, variation in diameter further up the bed being relatively small. For instance, Malek *et al.* [127] found that in 10 cm and 15 cm semi-circular columns, spout diameters measured at various levels starting from 2.5 cm above the orifice were generally within 10% of the mean value. Similar trends have been reported by other workers [114, 137, 152, 228, 259] for different column geometries. Several empirical equations correlating the mean value of spout diameter, based on measurements at various levels away from the immediate vicinity of the orifice, have been published. These equations, together with the range of experimental data supporting each one, are listed in Table 5.2. The most recent equation proposed by McNab [148] is based on a statistical regression analysis and gives a somewhat better fit of the data than the earlier equations, although the range of variables covered by it is no wider. The functional form of this equation is supported by the theoretical analysis of Bridgwater and Mathur [30] discussed in the following section. The McNab equation should therefore provide a sounder basis for estimating spout diameter in new situations than the other equations in Table 5.2.

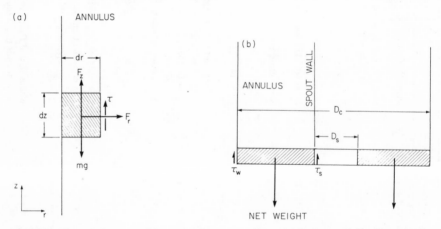

Fig. 5.2. Force balance models concerning spout shape. (a) Lefroy and Davidson [114]; (b) Bridgwater and Mathur [30].

C. Theoretical Models

Lefroy and Davidson [114], in an extension of the theoretical analysis discussed in Chapter 4, have attempted to explain the spout shape from a consideration of the balance of forces acting on an element of particles in the spout wall (see Fig. 5.2a). The weight of this element, according to their model, is supported by a combination of the upward force due to vertical flow of gas, F_z, and the vertical shear stress in the annular solids per unit area of spout wall, τ:

$$F_z + \tau = mg \tag{5.5}$$

The shear stress τ is considered to arise from each element being pressed outward against the neighboring element due to radial percolation of gas, and to be limited by the angle of repose ϕ of the material. Hence

$$\tau \leq F_r \tan \phi \tag{5.6}$$

F_z is calculable from the axial gas velocity distribution in the annulus

$$U_a = BU_{mf} \sin(\pi z/2H) \tag{3.22}$$

plus the condition that the drag force on the element at minimum fluidization will equal its weight. Assuming that the gas flow through the annulus is governed by Darcy's law, it follows that

$$F_z = Bmg \sin(\pi z/2H) \tag{5.7}$$

Equation (3.22), when combined with the gas balance

$$U_r = \left(\frac{D_c^2 - D_s^2}{4D_s}\right)\left(\frac{dU_a}{dz}\right) \tag{5.8}$$

gives

$$U_r = \frac{BU_{mf}\pi(D_c^2 - D_s^2)}{8D_s H} \cos\left(\frac{\pi z}{2H}\right) \tag{5.9}$$

and therefore, again assuming that Darcy's law applies,

$$F_r = \frac{Bmg\pi(D_c^2 - D_s^2)}{8D_s H} \cos\left(\frac{\pi z}{2H}\right) \tag{5.10}$$

The force balance of Eq. (5.5) can now be expressed as

$$\sin\left(\frac{\pi z}{2H}\right) + \frac{\pi(D_c^2 - D_s^2)}{8D_s H} \tan \phi \cos\left(\frac{\pi z}{2H}\right) \geq \frac{1}{B} \tag{5.11}$$

At $H = H_m$, $B = 1$; hence for a bed of maximum spoutable depth, Eq. (5.11) becomes

$$\frac{\pi(D_c^2 - D_s^2)}{8D_s H_m} \tan \phi \geq \frac{1 - \sin(\pi z/2H_m)}{\cos(\pi z/2H_m)} \tag{5.12}$$

From a comparison of the theoretical spout diameters required to satisfy the above equation against experimentally measured values, Lefroy and Davidson concluded that a constant spout diameter is possible for the upper half of the bed, but that in the lower half, the spout must taper down to less than half its upper diameter.

By combining the force balance Eq. (5.12) with the spout momentum balance Eqs. (4.14) and (4.15) and the particle entrainment Eq. (4.32), and introducing a number of approximations based on experimental observations, Lefroy and Davidson were able to derive the following dimensionally consistent expression for the constant value of spout diameter in the upper half of the bed:

$$D_s = 1.07 D_c^{2/3} d_p^{1/3} \tag{5.13}$$

An alternative force balance model, which takes into account solid stresses based on the principles of hopper flow of solids, has been proposed by Bridgwater and Mathur [30] for the middle region of the bed. The upper region gets excluded because solid stresses are taken to be independent of bed level—an assumption which is based on the behavior in hopper flow of solids and is valid only if the distance below the free surface is sufficiently large. The lowermost region is not covered, since effects due to gas entrance and the divergent walls of the conical base, as well as the normal stress exerted on the vessel wall due to radial percolation of gas (which occurs mostly in the lower part of the bed), are not taken into account. The model, depicted in Fig. 5.2b, is based on the postulates that the downward motion of particles at the vessel wall will lead to the exertion of an upward shear stress τ_w on the annular solids and that the movement of gas and particles in the spout will lead to an upward shear stress τ_s at the spout wall. The overall force balance becomes

$$\underbrace{\pi D_c \tau_w}_{\substack{\text{vessel wall} \\ \text{force}}} + \underbrace{\pi D_s \tau_s}_{\substack{\text{spout wall} \\ \text{force}}} = \underbrace{(\Omega \pi/4)(D_c^2 - D_s^2)}_{\text{effective weight}} \tag{5.14}$$

where Ω is the net downward force per unit volume exerted by the solids $= (\rho_s - \rho_f)(1 - \epsilon_a)g - dP_a/dz$, dP_a/dz being the vertical interstitial fluid pressure gradient in the annulus. The term τ_w was evaluated from a radial force balance on an element of the annular solids, assuming uniform lateral compression of the solids, while τ_s was expressed in terms of a friction factor between the fluid–particle suspension flowing in the spout and the spout wall, by analogy with pipe flow. Substitution in Eq. (5.14) then led to the following general equation for spout diameter:

$$\frac{32 f \rho_f Q_s^2}{\pi^2 \Omega (D_c - D_s) D_s^4} = 1 \tag{5.15}$$

where Q_s is the volumetric flow rate of fluid in the spout, and f the friction factor.

In order to obtain a working equation relating spout diameter with the independent variables of the system, a number of approximations based on experimental results had to be made by these workers too. The equation thus obtained, which is restricted to air-spouted beds, is as follows:

$$D_s = 0.0071 G^{1/2} D_c^{3/4} / \rho_b^{1/4} \tag{5.16}$$

where G is the air mass flow rate in lb/hr-ft^2, D_s and D_c are in feet, and ρ_b is in pounds per cubic foot. Comparison of the above equation with the empirical equation of McNab [Eq. (5.4)] shows not only that the correct variables have been included in Eq. (5.16), but also that the predicted dependence of spout diameter with each variable is also substantially correct. Absolute values of D_s predicted by Eq. (5.16) are about one-third of those by Eq. (5.4), but better agreement is hardly to be expected in view of the major simplifications which had to be made in translating the theoretical model into a working equation.

The outcome of both theoretical models outlined above is a semiempirical equation, Eqs. (5.13) and (5.16), respectively, relating spout diameter with the independent variables of the system. The former equation shows spout diameter to be independent of solids density, bed depth, and excess gas flow rate above minimum spouting, which is contrary to observation; nevertheless, the dependence of spout diameter on column diameter and particle size, and its independence of orifice size, indicated by the equation are qualitatively correct. Equation (5.16), at least in form, is closer to reality, since it includes solids density directly, and H, excess flow, and d_p through G, the minimum value of G for spouting being a function of both H and d_p. Bridgwater and Mathur consider that the development of a more rigorous theoretical equation might be possible if the analysis of solid stresses of their model could be linked to the fluid percolation and particle collision analyses of Lefroy and Davidson.

The only theoretical attempt concerned with spout shape in the more complex lowermost region of the bed has been made by Volpicelli *et al.* [252], who applied Helmholtz instability analysis for the growth of a disturbance at the interface between two flowing fluids to determine the stable spout shape. This approach to the problem arose out of their observation that disturbances developed at the bottom of the spout and moved upward like ripples. Their analysis led Volpicelli *et al.* to the conclusion that self-adjustment of spout diameter in the region above the orifice occurs in such a way that the lateral velocity of the downward moving annular solids in this region remains below a certain maximum value, which is similar to that obtained in gravity flow of solids through the bottom of a converging bin. On this basis, they predicted that pinching of the spout should occur at a short distance from the inlet orifice. From data obtained on the celerity of the wave disturbance from high speed motion pictures, the level at which spout pinching should occur was predicted to be 1–2 cm for their two-dimensional bed of 0.8 mm ceramic chips as against observed values of 4–8 cm (see Fig. 5.1e). An interesting corollary of the above analysis is the suggestion of Volpicelli *et al.* that gas leakage into the annulus is an effect but not the cause of spout instability. In partial support of this suggestion, they have cited the observation that spouted beds of any height could be operated in their column when disturbance growth was prevented by enclosing the spout inside a wall which was permeable to the gas but not to the solids.

5.2 VOIDAGE DISTRIBUTION

A. Annulus

The solids in the annulus of a spouted bed are essentially in the loose packed condition [58, 145]. The voidage in this region is therefore substantially constant and equal to that in a fixed bed of loosely packed particles. Minor variations, such as are known to occur in moving packed beds with solids flow rate [224] and gas percolation [218] are, however, bound to exist. Indeed, slight differences of voidage in different parts of the annulus are often observable by the naked eye, as noted by Thorley *et al.* [227], who attributed such differences for their wheat beds to particle orientation. These effects are, however, unlikely to be significant for most purposes.

B. Spout

The spout is like a riser through which particles are being transported in a dilute phase, with the added features of a decreasing gas flow and an increasing solids flow along the height. Spout voidage is close to 100%

immediately above the gas inlet orifice and decreases along the spout height. The voidage also varies radially across the spout, reaching a peak value at the spout axis.

1. *Measurement*

The vertical voidage profile was deduced by Mathur and Gishler [137] from certain observed data on gas and solids flow, using two different methods which are outlined below:

(a) The downward solids flow in the annulus, calculated from the particle velocity at the wall, should equal the upward flow in the spout at any bed level. Knowing the upward linear velocity of particles in the spout (measured by high speed cinephotography in a half column) and the spout diameter, the authors calculated the bulk density of the gas–solids suspension as a function of bed level, expressing the results in terms of voidage.

(b) As in the case of a vertical riser, the total pressure drop along the spout height was considered to be composed of (i) a solids static head equivalent to the dispersed solids bulk density, (ii) an acceleration pressure drop, and (iii) a solids friction loss due to relative motion of particles and gas with respect to the spout wall.

Thus,

$$dP_{total} = dP_{weight} + dP_{acceleration} + dP_{friction}$$

or

$$dP = dP_w + dP_{a+f} \tag{5.17}$$

From an energy balance over an increment of spout height dz,

$$dP_{a+f} = -(1/2\overline{U}_s A_s)\,d(m_s v_m{}^2) \tag{5.18}$$

where v_m is the radial mean upward particle velocity in the spout. Now

$$dP_w = d(\rho_{bs} z) \tag{5.19}$$

and

$$m_s = v_m A_s \rho_{bs} \tag{5.20}$$

where ρ_{bs} is the bulk density in the spout. Dividing Eq. (5.18) by Eq. (5.19) and combining the results with Eq. (5.20) gives

$$dP_{a+f}/dP_w = -[d(v_m{}^3 \rho_{bs})/d(\rho_{bs} z)][1/2\overline{U}_s] \tag{5.21}$$

Using the observed vertical profile of v_m (Figs. 4.2 and 4.3) and of $\overline{U}_s$ (estimated from the measured pressure drop profile as explained in Chapter 3), the authors evaluated the right-hand side of Eq. (5.21) for each 5 cm increment of the spout, assuming constant ρ_{bs} over this increment. This value, on substitution in Eq. (5.17) along with the measured total pressure drop for the increment, gave the corresponding pressure drop due to the solids bulk density in the spout. From the variation of bulk density with spout level, the voidage profile was calculated. The results thus obtained agreed well with those from method (a).

It should be pointed out, however, that Mathur and co-workers [137, 227, 228] in the early work at NRC measured the spout particle velocities from cine film without paying attention to radial variations. The velocity measured by them at a particular bed level would therefore be only approximately equal to the radial mean value, which is the velocity required for calculation of radial mean voidage by both the methods described above.

Direct measurements of spout voidage have been subsequently made using four different techniques:

(1) The piezoelectric technique mentioned in Chapter 4 has been used by some Soviet workers [153, 159]. Simultaneously with particle velocity measurements, they recorded the frequency with which the solid particles collided with the piezocrystal from the number of peaks observed on the oscilloscope per unit time. The local voidage at the probe tip was calculated from these data, using the following equation:

$$\epsilon_s = 1 - (\pi d_p^{\,3}/6)(N/vA_e) \tag{5.22}$$

where N is the number of collisions per second and A_e is the cross-sectional area of the sensing element.

(2) A capacitance probe was used by Goltsiker [78] for measuring local voidages in two-dimensional beds. Measurements by this method could be made not only in the spout, as with the piezoelectric probe, but in the entire bed, including the annulus and the interface region between the spout and the annulus.

(3) A β-ray absorption technique (described in Chapter 12) was employed by Elperin et al. [56] primarily to study the effect of pulsating the gas flow on spout voidage. Their data include some results for steady spouting also. The ability to measure voidage without disturbing the movement of particles by insertion of a probe is a major advantage of this method, but the practical difficulties in setting up a deep penetration β-ray system limited the experiments to a 12 mm thick two-dimensional bed.

(4) A photographic method was attempted by Lefroy [113]. His technique consisted in photographing the spout in a half-sectional column and comparing the number of particles per unit area with the number in a similar area of the packed bed annulus, for which the voidage was known. Lefroy

has, however, noted that the method was open to error, particularly at high voidages, since the depth of focus of the camera was only about 1 particle diameter, so that particles even slightly away from the column face would not give a sharp image on the photograph, and hence may not have been counted. The results obtained are nevertheless plausible when compared with those obtained by the more sophisticated techniques mentioned above. This would suggest that the error feared by Lefroy could not have been very large.

2. Experimental Results

Most of the available data for cylindrical columns are shown in Fig. 5.3 as vertical voidage profiles in the spout. The measurement method and the experimental conditions for each profile are tabulated in Table 5.3. The voidage value at a particular level represents the average voids over the spout cross section at that level. The upper limit of spout level covered by the data is the surface of the bed proper, no measurements having been made in the fountain above the bed. The data of Elperin *et al.* have been excluded since it is not clear from their paper what solid material the results refer to, although these results are presented later in the discussion on pulsed flow versus steady spouting in Chapter 12.

Based on the postulate that an upper limit on spoutable bed depth is set by the choking condition at the top of the spout, one might expect a decrease in ϵ_s at that level with the ratio H/H_m, and hence a progressively steeper voidage profile with increasing values of H/H_m. This expectation is to some extent borne out by the data in Fig. 5.3, which fall into two distinct groups. The flatter profiles shown in the upper part of the figure are all for $H/H_m < 0.5$, while the steeper profiles below are all for $H/H_m > 0.5$. The shapes of the individual curves, however, do not show any consistent trend with H/H_m. This is hardly surprising considering the variety of measurement techniques used, together with the fact that the choice of H/H_m as the governing parameter is open to question since mechanisms other than spout choking might also be responsible for determining the maximum spoutable depth, as discussed in the following chapter.

The general trend of spout voidage variation with bed level in wholly conical vessels has been found to be different from that in cylindrical columns. On the basis of results obtained in vessels of various cone angles and using several solid materials, Mukhlenov and Gorshtein [159] reached the conclusion that in a given bed, the spout voidage is substantially constant over the height of the spout. This conclusion obviously does not apply to the lowermost part of the spout, over which the voidage would decrease from 100% at $z = 0$ to its constant value. The height of this lowermost part is

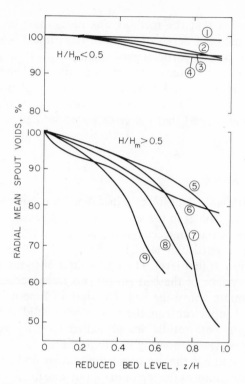

Fig. 5.3. Measured vertical voidage profiles in the spout (see Table 5.3).

reported by Goltsiker [78] as 4 cm in a 11 cm deep bed contained in a conical vessel of 40° angle; above the 4 cm level, Goltsiker too found the spout voidage to remain substantially unchanged with bed level. An empirical correlation, based on dimensional analysis, of their spout voidage values has been proposed by Mukhlenov and Gorshtein:

$$\epsilon_s = 2.17(\mathrm{Re_i}/\mathrm{Ar})^{0.33}(H/D_i)^{-0.5}(\tan\theta/2)^{-0.6} \qquad (5.23)$$

The data supporting Eq. (5.23) cover Reynolds numbers (based on air velocity through the orifice and diameter of the particle) from 50 to 1100, Archimedes numbers from 6.27×10^4 to 21.25×10^4, and H/D_i from 1 to 9, and included cone angles from 20° to 60°.

Radial profiles of local voidage within the spout have been measured in conical beds by Elperin *et al.* [56] and by Romankov, Goltsiker, and others [78, 202]. Typical data for the two groups of investigators, presented in Fig. 12.7 and Fig. 5.4, respectively, show the existence of a roughly parabolic profile with a maximum value at the spout axis in both cases. There is, however, one significant point of difference. The profile for Fig.

TABLE 5.3 Experimental Methods and Conditions Corresponding to Spout Voidage Data of Fig. 5.3

Curve number	Measurement method	Material	d_p (mm)	ρ_s (Mg/m³)	D_c (cm)	D_i (cm)	H (cm)	H/H_m	Ref.
1	Piezoelectric	Millet	2.0	1.18	9.4	1.5	10	0.33[a]	[153]
2	Indirect	Wheat	3.6	1.41	61	10	122	0.26[a]	[227]
3	Piezoelectric	Wheat	3.6	1.41	9.4	1.5	10	0.44[a]	[153]
4	Piezoelectric	Silica gel	4.0	1.12	9.4	1.5	10	0.43[a]	[153]
5	Indirect	Wheat	3.2	1.41	15	1.27	64	0.84	[137]
6	Piezoelectric	Silica gel	6.0	1.12	9.4	1.5	12	0.70[a]	[153]
7	Photographic	Kale seeds	1.74	1.0	30.5	3.5	84	0.69	[113]
8	Photographic	Polyethylene chips	3.5	0.89	30.5	3.5	100	0.91	[113]
9	Photographic	Kales seeds	1.74	1.0	30.5	3.5	122	1.0	[113]

[a] H_m for these systems was calculated from Eq. (6.5); measured values of H_m were used for the others.

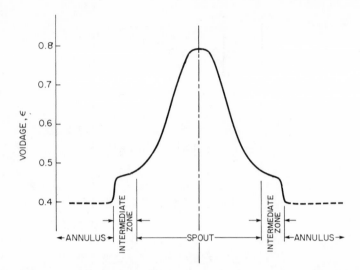

Fig. 5.4. Radial voidage profile in a conical spouted bed measured by Goltsiker [201, p. 63].

5.4 shows a voidage plateau at the spout periphery. This narrow region, which has a voidage (45–50%) intermediate between the dense phase annulus and the dilute phase spout core, has been identified by Romankov *et al.* as the third zone (in addition to the annulus and the spout core), comprising "particles descending with a quick vortex-like movement," and can be seen in Fig. 1.2b. According to these authors, net downward particle velocities in the intermediate zone are several times higher than in the surrounding annular zone. No such intermediate zone is discernible in the profile of Fig. 12.7 (p. 254), nor has such a zone been observed by other investigators in cylindrical beds.

3. *Theory*

From a theoretical point of view, spout voidage is part and parcel of the complete bed hydrodynamics, including both gas and solids flow patterns as well as spout shape. Any analysis of voidage distribution therefore must of necessity be all-inclusive. The only such analysis attempted so far is that due to Lefroy and Davidson [114], which has been presented mainly in the context of solids flow but which reappears in one form or another in connection with several other aspects of spouted bed behavior, such as pressure drop, gas flow pattern, spout diameter, and now spout voidage.

Referring to Chapter 4, we find that numerical integration of Eqs. (4.18) and (4.19) in conjunction with Eqs. (3.21) and (4.21) yields vertical profiles

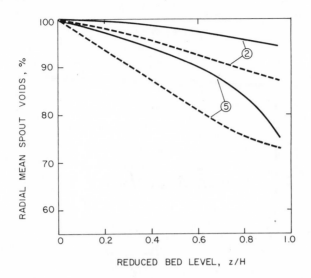

Fig. 5.5. Vertical spout voidage profiles: Theory versus experiment. - - - Theoretical; ② B in Eq. (4.21) taken as 0.6, $D_s = 8.1$ cm for all z/H; ⑤ B taken as 0.95, $D_s = 2.26$ cm for all z/H; —— experimental. Curve numbers correspond to those in Fig. 5.3 and Table 5.3.

not only of gas and particle velocities but also of spout voidage. Although Lefroy and Davidson have presented computed voidage profiles for their beds of kale seeds and polyethylene chips, they did not compare the computed results against measured values since they were obviously not satisfied with the reliability of their photographic method for voidage measurements. Our own calculations of particle velocities in wheat beds represented in Figs. 4.2 and 4.3 yielded theoretical voidage profiles too, corresponding to experimental curves numbered 2 and 5, respectively, in Fig. 5.3 (and Table 5.3). The comparison between theory and experiment in Fig. 5.5 shows that even for the larger bed (curve 2), which is free from the complication of a variable spout diameter, the theoretical profile is considerably steeper than the experimental profile. The same general trend exists for the smaller bed, although here the two curves are quite different in shape also. Calculated results for the latter bed, allowing the spout to expand up to $z/H = 0.33$ (i.e., corresponding to curve 2 in Fig. 4.3) are not included in Fig. 5.5, but these showed even poorer agreement with experimental values, the predicted voidages being greater than 0.97 at all bed levels. The discrepancies noted above are consistent with those which emerged from the particle velocity comparison in Chapter 4, and must arise from the limitations of the theory and the difficulties in testing its validity already discussed in Chapter 4.

6 | *Spouting Stability*

It was explained in Chapter 1 that the regime of stable spouting is critically dependent on certain conditions; unless these are satisfied, the movement of solids becomes random, leading to a state of aggregative fluidization, and with increase in gas flow, to slugging. Spouting can be achieved only within certain limits of solids properties, while whether or not a material having properties within these limits will spout depends upon the geometry of the column, including to some extent the design of the gas inlet. A further overriding restriction on spouting stability is imposed by bed depth, since spouting action for any given solids properties and column geometry would terminate beyond a certain maximum depth. The maximum spoutable depth can therefore be looked upon as an index of spouting stability, although a stably spouting bed of a depth smaller than the maximum would become unstable at excessively high gas flow rates.

The first part of this chapter briefly summarizes experimental observations concerning the effect on spouting stability of the various factors involved, while the second part deals with methods for predicting maximum spoutable bed depth.

112

6.1 EXPERIMENTAL OBSERVATIONS

A. Effect of Column Geometry

1. *Orifice to Column Diameter Ratio*

In a given column, the maximum spoutable bed depth H_m decreases with increasing orifice diameter until a limiting value of D_i is reached beyond which spouting no longer occurs (see Fig. 6.1). On the basis of his data for spouting of several materials in cylindrical columns, Becker [15] suggested that the critical value of the ratio D_i/D_c equals 0.35. While this value is approximately in line with the data for wheat shown in Fig. 6.1, the critical value for finer particles is considerably smaller, being 0.1 for sand of $d_p = 0.6$ mm (see Fig. 1.5a, b). An analysis of the abovementioned data for wheat and sand led Németh and Pallai [166] to suggest that there is interaction between the limiting value of D_i/D_c and the ratio D_c/d_p, and that the limiting value reaches a maximum (~ 0.3) for $D_c/d_p = 40$–50.

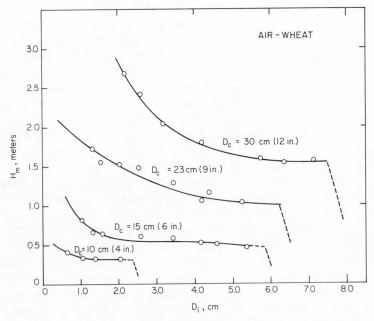

Fig. 6.1. Effect of orifice diameter on maximum spoutable bed depth [137].

For conical vessels, the existence of an upper limit to the ratio of orifice to bed-surface diameter has been established by Romankov and Rashkovskaya [201, p. 47], who have shown this ratio to be dependent on the Archimedes number for materials of 0.36–9 mm diameter.

2. *Cone Angle*

The lower conical section of the bed facilitates the flow of solids from the annulus into the gas jet region. With a flat instead of a conical base, a zone of stagnant solids with a conelike inner boundary is formed at the base, but this does not affect spouting stability. If the cone is too steep, on the other hand, spouting becomes unstable since the entire bed tends to be lifted up by the gas jet. This applies equally to cylindrical vessels with a conical base and to entirely conical vessels. The limiting cone angle depends to some extent on the internal friction characteristics of the solids, but for most materials its value appears to be in the region of 40° [55; 94; 201, p. 64].

3. *Inlet Design*

In the early work at the National Research Council of Canada, it was found by trial and error that spouting was more stable when the gas inlet orifice was somewhat smaller than the narrow end of the cone (see Fig. 6.2a). This finding was subsequently rationalized by Manurung's [134] demonstration that maximum stability is obtained with a design which does not permit the gas jet to be deflected from the vertical path before it enters the bed of particles. In his own experiments, he achieved this end by using the design shown in Fig. 6.2b, the main feature of which is that the gas inlet pipe protrudes a short distance above the flange surface. With this inlet, Manurung obtained somewhat higher maximum spoutable depths for several materials and was able to achieve stable spouting for coal beds containing a high proportion of fines, which would not spout with other gas inlets in which the gas pipe did not protrude. The stabilizing effect of a slightly protruding inlet pipe has been confirmed by Reddy *et al.* [192], who obtained better results with a converging nozzle inlet (Fig. 6.2c) than with a straight pipe. These workers consider that the flat section or ledge between the inlet nozzle and the lower end of the cone plays an important role in stabilizing the spouting flow pattern. Although the inlet of Fig. 6.2c gave a maximum spoutable depth only slightly higher than those with the other types of inlets tried, the spouting action at any bed depth was more stable with this particular inlet. Malek and Lu [129] did not observe any effect on H_m of their two inlet designs, but did find that the exact positioning of the screen had a marked effect on spouting stability. If the screen was

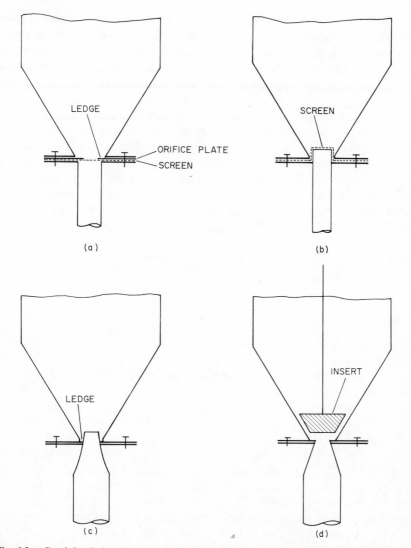

Fig. 6.2. Gas inlet designs for improved stability: (a) Constricting orifice; (b) straight inlet pipe, protruding; (c) converging nozzle, protruding; (d) truncated conical plug insert.

"loosely fitted" over the orifice plate, spouting at any depth was unstable, but when the screen was placed below the orifice plate (as in Fig. 6.2a), satisfactory spouting resulted. The use of a converging–diverging gas inlet pipe has been mentioned by Berquin [20], although it is not clear how spouting stability would be affected by such a design.

A curious method of improving stability discovered by Peterson [180] is the use of an insert located a few centimeters above the gas inlet opening, as shown in Fig. 6.2d. In the case of a 61 cm diameter × 122 cm deep wheat bed with a 12.7 cm diameter gas inlet, such an insert caused a dramatic improvement in spouting stability and a substantial reduction (20%) in the minimum air flow rate required for spouting.

From the few observations cited above, it is clear that the exact design of the gas inlet has an important effect on spouting stability. The question of inlet construction has, however, received insufficient attention. Indeed, it is even possible that some of the discrepancies in other aspects of spouting behavior observed by different investigators may be due to unspecified differences in inlet designs. Factors other than spouting stability which would enter the design of the gas inlet arrangement are discussed in Chapter 13.

B. Effect of Solids Properties

1. *Particle Size*

Although the minimum particle size for spouted bed operation has been quoted as 1 mm in Chapter 1, Ghosh [75] has suggested that spouting action can be achieved for much finer materials, as long as the gas inlet diameter does not exceed 30 times the particle diameter. Using a very small air orifice, he was able to obtain a miniature spouted bed ($\sim$ 1 cm diameter) with glass beads as fine as 80–100 mesh (mean diameter = 0.16 mm). Spouting of such fine material cannot, however, be achieved on a larger scale, except perhaps by the use of a bed with multiple spouts in parallel (see Chapter 12). If the column diameter is increased without a corresponding increase in orifice diameter, the spouting action remains confined to a small region near the orifice, leaving the surrounding solids stagnant. The entire bed can be made to circulate by increasing the orifice size also, but in a less organized fashion than in a true spouted bed of coarse particles. Such bed behavior has been called "nonhomogeneous fluidization" by Vainberg *et al.* [241], who worked with conical vessels and with particles ranging in size from 0.18 mm to 1 mm. Baskakov and co-workers [11, 12], however, refer to their beds of similar size solids as spouting beds, no doubt using the term spouting in a loose sense.

The maximum spoutable bed depth was found to decrease with increasing particle size by Malek and Lu [129], who experimented with four different sizes of wheat (1.2–3.7 mm) in a 15 cm diameter column. On the other hand, Reddy *et al.* [192], who worked with mixed size materials (alundum, glass spheres, and polystyrene), also in a 15 cm column, reported that H_m first increases with particle size and then decreases, a peak value being attained at a mean particle size of 1.0–1.5 mm. The observed variation of H_m, correlated by Reddy *et al.* with mean particle size, is likely to be also influenced by size distribution, which cannot be fully characterized by any particular mean diameter. Nevertheless, the existence of a peak value of H_m with respect to particle size alone is theoretically predictable from a comparison of the effect of particle size on gas velocities required for spouting and for fluidizing a given material. From Eq. (2.38), the effect of particle size and bed depth on spouting velocity, with all other variables held constant, is as follows:

$$U_{ms} = kd_p H^{1/2} \qquad (6.1)$$

while the general dependence of fluidization velocity on particle size can be expressed in the following form:

$$U_{mf} = Kd_p{}^n \qquad (6.2)$$

Since at H_m, $U_{ms} \simeq U_{mf}$, it follows from Eqs. (6.1) and (6.2) that

$$H_m \simeq K' d_p^{(2n-2)} \qquad (6.3)$$

Now the value of n in Eq. (6.3) depends on the flow regime. Using either the generalized equation of Wen and Yu [258] cited by Reddy *et al.* and analyzed by Kunii and Levenspiel [110, p. 73], or the Richardson–Zaki equation [194], in conjunction with the terminal velocity laws of Stokes and Newton, it can be shown that n changes from 2 for laminar flow to 0.5 for turbulent flow. Hence the exponent on d_p in Eq. (6.3) would change with increasing d_p from a value of 2 to -1, depending upon the flow regime, causing a peak value of H_m to occur with respect to particle size—at $n = 1$.

For the mixed size materials used by Reddy *et al.*, the peak value of H_m was found to occur at a mean particle size of 1.0–1.5 mm, the corresponding value of Reynolds number being about 70 in all cases, regardless of the particle density or the orifice size used (see Fig. 6.3). The experimental data in Fig. 6.3 follow slopes of 1.0–2.0 for the ascending section and of -0.4 to -0.8 for the descending section, which includes the data of Malek and Lu for different sizes of wheat particles. Thus the range of experimental slopes are similar to those predicted, despite the possible influence of size distribution in the data of Reddy *et al.*, mentioned above.

2. Size Distribution

Uniformity of particle size favors spouting stability, since the lower permeability of a bed containing a range of sizes would tend to more effec-

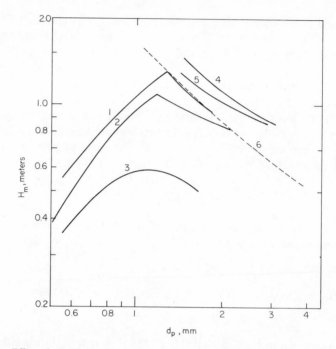

Fig. 6.3. Effect of particle size on maximum spoutable bed depth in a 15 cm diameter column. Data of: ——Reddy *et al.* [192]; ——— Malek and Lu [129]. Curve 1: alundum, $D_i = 9.5$ mm; 2: alundum, $D_i = 12.7$ mm; 3: glass spheres, $D_i = 9.5$ mm; 4: polystyrene, $D_i = 9.5$ mm; 5: polystyrene, $D_i = 19.0$ mm; 6: wheat, $D_i = 25.4$ mm.

tively distribute the gas rather than produce a jet action. The presence of a small proportion of fines in a closely sized bed can seriously impair spoutability ([103], $-40 +60$ mesh sand in a bed of $-20 +30$ mesh sand), while the addition of a small proportion of coarse particles to a bed of finer material can also have the same effect ([219], $-9 +14$ mesh alundum added to a bed of $-35 +48$ mesh alundum). The limits of particle size spread beyond which spouting would no longer occur are nevertheless fairly wide, the latitude being greater with large particles than with small. Thus, beds of wood chips and cellulose acetate containing up to an eightfold size spread with particles of up to 3 cm in size could be satisfactorily spouted [43], while for coal and alundum with maximum size in the 2–4 mm range, limiting size spreads were seldom more than fivefold [134, 219]. Evidence from the comminution experiments described in Chapter 12, however, shows that the stability of a bed of ~5 mm diameter uniform size particles is less sensitive to the presence of very much finer material than suggested by the above observations on beds containing a spectrum of particle sizes.

3. Density

Solids with widely differing densities, ranging from wood chips to iron pellets, have been spouted without any indication that any limits of particle density exist beyond which spouting action would not be achieved. Nor is there any clear evidence to show whether spouting stability is affected by particle density or not. The empirical correlation for calculating H_m proposed by Malek and Lu [129], Eq. (6.5), implies that spouting stability is adversely affected by particle density, but Fleming [63] was able to spout deeper beds of alundum particles ($\rho_s = 2.46$ Mg/m^3) than of glass beads ($\rho_s = 1.55-1.85$ Mg/m^3) of the same size. The above observations are, however, confused by the effects of particle shape and surface characteristics, which were certainly different for the different materials mentioned.

4. Particle Shape and Surface Characteristics

These two factors undoubtedly have a marked influence on spouting stability but their effect has proved difficult to evaluate, partly because the shape and especially the surface characteristics of solid particles are not easy to define. Using angle of repose as a combined criterion for both shape and surface (irregular and rough particles = high angle of repose), Fleming [63] noted a direct dependence between angle of repose and spouting stability in his results for alundum, polystyrene, and glass beads. The empirical equation of Malek and Lu [129], Eq. (6.5), also suggests that deeper beds of nonspherical particles can be spouted than of spherical particles. However, in beds of certain particles which deviate very widely from the spherical, true spouting action seems to terminate altogether. Thus, in the case of strongly ellipsoidal particles such as flaxseed and barley, Becker [15] observed that although a through channel resembling a spout was formed by the air jet in the bed, the resulting agitation of the solids was much more feeble than in true spouting. The "pseudo-spout" formed was insensitive to changes in inlet to column diameter ratio and bed depth, and behaved like a solids free channel, which probably owed its stability to the interlocking tendency of the particles. A similar phenomenon was observed by Reddy *et al.* [192] in their experiments using deep beds of polystyrene.

C. Effect of Gas Flow Rate

In relatively shallow beds ($H/D_c < 3$), an increase in gas flow much above that required for minimum spouting causes the spout above the bed surface to lose its well-defined shape, and though the movement of solids in the region above the bed becomes chaotic, the regular downward motion of particles in the annulus remains intact [137]. In deeper beds, on the other hand, the solids movement in the bed itself is disrupted at high flow rates. This disruption in the case of coarse particles takes the form of slugging, while with fine materials the bed first passes into the fluidized state and with further increase in flow, to slugging. Phase diagram data similar to that of Fig. 1.5, reported for beds of various materials in 15 cm diameter [134, 137] and 23 cm diameter [49] columns, suggest that, in general, spouting stability with respect to gas flow rate increases with increasing particle size, increasing column diameter, decreasing orifice to column diameter ratio, and decreasing bed depth. For a given material and column geometry, the range of permissible gas flow for stable spouting becomes narrower as the bed depth approaches the maximum spoutable under those conditions. No data on the upper gas flow limits with larger columns have been reported, but since bed depths used in large columns would normally be well below the maximum spoutable, the tolerance for excess gas in such columns should be large.

6.2 MAXIMUM SPOUTABLE BED DEPTH

A. Controlling Mechanisms

There are three distinct mechanisms which could conceivably cause spouting to become unstable beyond a certain upper limit of bed depth.

1. *Fluidization of Annular Solids*

Radial percolation of gas from the spout into the annulus causes the upward gas velocity in the annulus to increase along the height of the bed, as discussed in Chapter 3. It stands to reason that if the annular gas velocity becomes high enough to fluidize the uppermost layer of solids in a bed of depth H, the smooth downward motion of annular solids would be disrupted and the spouting action would give way to either aggregative fluidization or slugging, thus making $H = H_m$.

2. *Choking of the Spout*

Alternatively, the transition from spouting to slugging can originate within the spout by the same mechanism which causes slugging in an overloaded riser [252].

3. *Instability Growth*

The spout is more analogous to a wave disturbance than to a riser, since its outer boundary is not a rigid wall but rather a hydrodynamically induced interface. Indeed, the spout–annulus interface displays an observable rippling movement, which is particularly pronounced in the lower region of the bed. Volpicelli *et al.* [252] have postulated that it is the growth of this surface instability which may cause spout disruption and hence control the maximum spoutable bed depth.

It is likely that under a given set of hydrodynamic conditions, the onset of slugging is brought about by one or the other of the three mechanisms outlined above, but these conditions have not been selectively determined or even identified.

B. Prediction Methods

In developing working equations for predicting H_m, the latter two mechanisms have been ignored altogether, the correlations proposed being based either on the first mechanism or simply on empiricism.

If one assumes that it is the fluidization of the upper layer of annular solids which causes spouting action to terminate at a certain bed height, H_m, the gas velocity through the annulus at $z = H = H_m$ must equal U_{mf}. Gas velocity through the spout $\overline{U}_s$ at the same level would be higher than through the annulus (see Fig. 3.1) but since the cross-sectional area of the spout is always a small fraction of the annulus area, U_{ms} at H_m (i.e., U_m) should approximately equal U_{mf}. Experimentally, the above equality has been found to hold under certain conditions only and is by no means universal, measured values of U_m being up to 50% higher than U_{mf} as discussed in Section 2.3.A. Thus it would appear that for cases where U_m is considerably greater than U_{mf}, termination of spouting with increasing bed depth is actually caused not by fluidization of the annular solids but by one of the other two mechanisms mentioned in the preceding section. Nevertheless, Thorley *et al.* [228] were able to predict values of H_m, though only approximately, under a variety of conditions by simultaneously solving an equation for U_{ms} with an equation for U_{mf}. This approach was subsequently adopted by other workers with variations in the particular equations used for calculating spouting and fluidization velocities, which are all listed in Table 6.1. It should be noted that Becker [15], unlike the others, did not rely on the assumption that $U_m = U_{mf}$. Instead, he empirically derived Eq. (2.35) from specific data for U_m, although this equation did not lead to any net improvement over the simpler approach because of shortcomings in other respects, as discussed in Chapter 2. Of the direct empirical correlations in Table 6.1,

TABLE 6.1   Empirical Equations for Prediction of H_m

Investigators	Equations

Indirect method involving simultaneous solution of equations for U_{ms} and U_m

Thorley *et al.* [228]	U_{ms}	from Eq. (2.38), Mathur and Gishler equation
	$U_m \sim U_{mf}$	from Eq. (2.36), based on Ergun equation
Becker [15]	U_{ms}	from Eq. (2.44)
	U_m	from Eq. (2.35)
Manurung [134]	U_{ms}	from Eq. (2.45)
	$U_m \sim U_{mf}$	from Eq. (2.36)
Reddy *et al.* [192]	U_{ms}	from Eq. (2.46)
	$U_m \sim U_{mf}$	from Wen and Yu equation [258]

Direct empirical correlations

Reddy *et al.* [192]

$$H_m = 11.6 d_v^{1.26} D_i^{-0.33} \rho_s^{-0.2} \quad \text{for} \quad \text{Re} \leq 70 \quad (6.4a)$$
$$H_m = 20.4 d_v^{-0.57} D_i^{-0.125} \rho_s^{-0.2} \quad \text{for} \quad \text{Re} \geq 70 \quad (6.4b)$$

with H_m, d_v, and D_i in inches and ρ_s in pounds per cubic foot. Supporting data cover only 1 column diameter (15 cm) but several materials of mixed size in the range 0.25–3.3 mm.

Malek and Lu [129]

$$H_m/D_c = 0.105(D_c/d_p)^{0.75}(D_c/D_i)^{0.4}(\lambda^2/\rho_s^{1.2}) \quad (6.5)$$

with ρ_s in megagrams per cubic meter. λ is a particle shape factor (see Appendix) with values ranging from 1.0 for millet, sand, and timothy seed to 1.65 for gravel. Data, correlated within $\pm 11\%$, cover different column diameters (10–23 cm), several uniform size materials ($d_p = 0.8$–3.7 mm, $\rho_s = 0.91$–2.66 Mg/m^3), and different orifice sizes.

only the Malek and Lu [129] equation is supported by a sufficiently wide range of variables to be of practical interest. The general applicability of this equation is discussed later.

The only theoretically based attempt to derive an expression for H_m has been made by Lefroy and Davidson [114], as an extension to their analysis of the mechanics of spouted beds discussed in previous chapters. From measurements of spout diameter made in beds of kale seeds, polyethylene chips, and peas at maximum spoutable depths, they determined the value of the left-hand side of their force balance equation [Eq. (5.12)] and found it to be about 0.36 in all cases. Therefore

$$\frac{\pi(D_c^2 - D_s^2)}{8 D_s H_m} \tan \phi = 0.36 \quad (6.6)$$

Assuming that $D_c \gg D_s$, and taking a constant value of $\phi = 33°$ for all three materials, Eq. (6.6) simplifies to

$$H_m = 0.72D_c^2/D_s \tag{6.7}$$

This equation can be combined with Eq. (5.13) to yield the following dimensionally consistent expression for H_m in terms of primary parameters:

$$H_m = 0.67D_c^{4/3}/d_p^{1/3} \tag{6.8}$$

It should be noted that in common with the indirect methods for estimating H_m listed in Table 6.1, Eq. (6.8) is also ultimately based on the assumption that disruption of spouting is caused by fluidization of the annular solids. This assumption enters via Eq. (5.7), which was used in the derivation of Eq. (5.12).

Although Eq. (6.8) was found to be in approximate agreement with the experimental data on which it was based, it does not, again in common with the more empirical methods, encompass the observation of previous workers that H_m first increases before it starts to decrease with increasing particle size. This limitation of Eq. (6.8) has been attributed by Lefroy and Davidson to the breakdown in the case of very fine particles of the assumption of uniform pressure across a horizontal section of the annulus, which is essential to the derivation of Eq. (6.7). To take into account the effect of inlet diameter, which is missing in Eq. (6.8), another semiempirical equation for H_m was derived, though implicitly [by combining Eqs. (31) and (32) of their paper], on the basis that the momentum transfer rate at the spout inlet is observed to be roughly one-half of the total upward force necessary to support the bed at minimum spouting:

$$H_m = 0.192d_p D_c^4/D_i^2 D_s^2 \tag{6.9}$$

However, when Eq. (6.9) is combined with Eq. (5.13) to eliminate the dependent variable D_s, the expression obtained,

$$H_m = 0.168d_p^{1/3} D_c^{8/3}/D_i^2 \tag{6.10}$$

shows the effect of d_p on H_m to be contradictory to that given by Eq. (6.8) and that of D_c to be considerably different. The conflict between Eqs. (6.8) and (6.10) arises partly from the weakness of the assumption that the inlet gas momentum is a fixed proportion of the overall pressure drop, used in arriving at Eq. (6.9) [115]. It is also likely that other assumptions made by Lefroy and Davidson in developing their various models are not all mutually compatible [59]. Of their two predictive equations, Eq. (6.10) is more consistent with the observed variation of H_m with D_c represented by the empirical equation (6.5). With respect to the effect of d_p, the direct dependence of H_m observed for small particles (see Fig. 6.3) is correctly predicted by Eq. (6.10) and the inverse dependence observed for coarse particles [Fig. 6.3, also Eq. (6.5)] by Eq. (6.8), although the agreement in both cases is no better than qualitative.

An attempt by the present authors [142] to decide which of the various calculation methods proposed are suitable for predictive purposes showed Becker's method to be the most reliable when tested against a wide range of available experimental data covering different materials, column diameters, and air inlet sizes. The equations of Thorley et al., as well as the simple correlation of Malek and Lu, also gave fairly good predictions, though neither one was as consistently reliable as Becker's. The equations of Reddy et al. and of Lefroy and Davidson proved to be unsatisfactory for quantitative predictions. The conclusions above apply to solids of uniform size. Calculations were also performed for beds of mixed size particles using different mean diameters to characterize the mixture, but none of the equations appeared to have general validity regardless of which mean diameter was used.

Finally, it should be stressed that since the equations in Table 6.1 have all been arrived at without adequate consideration of the physical mechanism responsible for instability, it is doubtful if even the preferred equations mentioned above would extrapolate to conditions which are very different from those covered by the data used in testing their validity.

7 | *Particle Attrition*

The solid phase in a spouted bed is subjected to a certain amount of wear and tear during its passage through the high velocity spout region. The upward moving particles collide against each other within the spout and against the layer of annulus particles constituting the spout wall, as discussed in Chapter 4. Materials such as wheat and other grains, plastic granules, etc. are to some extent resilient and are therefore able to withstand this rough treatment without breaking down. With more fragile solids, particle attrition in a spouted bed can be considerable.

From the observations summarized in Table 7.1, it is seen that attrition is not always undesirable; indeed this feature of spouting becomes a distinct advantage for processing solids which tend to agglomerate, as in the case of coal carbonization and iron ore reduction. In shale pyrolysis, too, attrition plays a beneficial role in exposing fresh particle surface for retorting. Drying of pastes in a bed of glass beads and grinding of particulate solids, described in Chapter 11, are further instances where the comminutive action in the spout has been deliberately aggravated to good advantage. On the other hand, particle attrition can rule out the use of a spouted bed in certain cases, depending on the characteristics of the solid

TABLE 7.1 Some Observations concerning Particle Attrition

Bed solids	Type and scale of operation	Observations	Ref
Peas, flax, and lentils	Continuous drying in industrial unit $D_c = 61$ cm, $H = 214$ cm, $D_i = 8.9$ cm	Loose hulls (on peas), dirt, and mold escaped overhead. Dusting rate estimated as 1% or less of throughput. Spouting appears to be no more violent than bulk-handling procedures.	Peterson [178] (see Table 11.2)
Cellulose acetate, matted fibrous particles mostly in the 2.5–20 mm size range; moisture 30% dry basis	Batch tests using unheated air in a half-round column $D_c = 61$ cm, $H \sim 240$ cm, $D_i = 5$ cm	After 1 hr of spouting, 9.5% of initial bed weight became finer than 2.5 mm. Spouted bed drying ruled out because of excessive attrition.	Cowan et al. [43]
Wood chips, size spread similar to above; moisture 50% dry basis	Batch tests using unheated air in a half-round column $D_c = 61$ cm, $H \sim 130$ cm, $D_i = 10$ cm	After 1 hr, fines (<2.4 mm) produced amounted to 2.3% of bed weight.	Cowan et al. [42]
Polyvinyl chloride (PVC), $d_p = 0.4$ mm Polyvinyl formal (PVF), $d_p = 1.8$ mm Colcothar, $d_p = 0.7$ mm	Continuous drying on the bench scale $D_c = 22$ cm, $H = 17$–28 cm, $D_i = 7.9$ cm	Loss by elutriation as weight percent of drier throughput was 5–10% for PVC, 3% for PVF, and 5–7% for Colcothar.	Auf et al. [5] (see Table 11.2)
Cement clinker, −6 + 10 mesh	Batch test $D_c = 15$ cm, $H = 50$ cm, $D_i = 1.27$ cm	After 1 hr of spouting, 10.5% of initial bed weight was elutriated out of the bed as fine dust (−65 + 200 mesh).	Mathur [143]

Material	Operation / Conditions	Observations	Reference
Ammonium nitrate, −10 +35 mesh	Continuous drying in industrial unit $D_c = 61$ cm, $H = 117$ cm, $D_i = 7.6$ cm	Collection in overhead cyclone during normal operation did not exceed 10% of drier throughput.	Indian Explosives [96] (see Table 11.2)
Coal, size range 1.0–3.3 mm	Batch tests $D_c = 15$ cm, $H = 63, 76$ cm, $D_i = 1.27$ cm	After $\frac{1}{4}$ hr, dust collected was 1% of bed weight. Attrition rates were higher for coarser bed material.	Manurung [134]
Coal, size range 1.6–3.3 mm	Continuous carbonization experiments with bed temperatures of 450–650°C $D_c = 15$ cm	Attrition had a beneficial effect since agglomerates of char particles were broken up.	Barton et al. [9] (see Chapter 11)
Iron ore and ore–flux pellets, size range 0.25–3 mm	Lab. scale reduction experiments with bed temperatures of 700–1000°C. $D_c = 2.5$ cm, $H = 1.5$–2.0 cm, $D_i = 0.8$–1.0 cm	Here again, attrition of agglomerates in the spout enabled reduction of the ore to be carried out at high temperatures without the problem of stickiness.	Vavilov et al. [245] (see Chapter 11)
Oil shale, up to 6 mm size	Pyrolysis at 510–732°C on the bench scale, continuous operation. $D_c = 16.5$ mm, tapering down to 5.1 cm; $D_i = 1.27$ cm	Attrition helped the process since outer surface of particle became fragile on loss of organic matter and was broken off in the spout, exposing fresh surface for retorting.	Berti [22] (see Chapter 11)
Pharmaceutical tablets	Batch coating operation lasting 60–90 min, industrial scale. $D_c = 38$ cm, batch size = 100 kg	Attrition and breakage problem solved to the point that no inspection or sorting is required. Soft or friable tablets are reformulated or seal-coated by other means prior to spouting.	Singiser et al. [217] (see Chapter 11)

TABLE 7.2 Size Reduction of Bed Solids during Spouting[a] of Urea Fertilizer (15 cm diameter × 35 cm deep bed)[b]

Experimental conditions	Duration of spouting (min)	Sieve analysis of bed, weight %					
		30° Cone			60° Cone		
		$-6 +9$[c]	$-9 +32$	$-32 +100$	$-6 +9$	$-9 +32$	$-32 +100$
$D_i = 25.4$ mm, $u_i = 31.7$ m/sec	0	100	0	0	100	0	0
	10	95	5	0	98	2	0
	30	95	5	0	98	2	0
	60	95	5	0	98	2	0
$D_i = 12.7$ mm, $u_i = 123$ m/sec	0	100	0	0	100	0	0
	10	91	8	1	96	4	0
	30	75	24	1	93	7	0
	60	71	28	1	91	9	0
$D_i = 6.3$ mm, $u_i = 394$ m/sec	0	100	0	0	100	0	0
	10	44	28	28	77	12	11
	25	44	28	28	64	24	12
	40	44	28	28	42	45	13

[a] See Haji-Zainali [88].
[b] Air flows used were slightly above those required for minimum spouting.
[c] Tyler mesh.

material involved and the extent of size reduction which can be accepted for the particular process. Thus in a catalytic reactor, attrition poses a more serious problem than in solids processing, since here the cumulative attrition over a prolonged period must remain within acceptable limits. The catalyst particles, therefore, need to be exceptionally rugged. It is probably this requirement, more than any other cause, that has impeded the development of the spouted bed as a catalytic reactor.

Apart from the fragility of the solid material, the extent of attrition would also depend on spouting conditions; it should therefore be possible to exercise some measure of control on attrition for a given material by the choice of design parameters. No detailed investigation into this important aspect of spouting has so far been reported, but some indication of attrition behavior is provided by experiments carried out at the University of British Columbia with beds of urea fertilizer in a 15 cm diameter column [209, 88]. Different inlet orifice sizes, cone angles, bed depths, and air flow rates were used, and sieve analyses of samples drawn from the spouting bed at 10-min intervals were determined. The quantity of fine dust collected in an overhead cyclone after 2 hours of spouting was also measured but this amounted to only 0–9 gm per kilogram of the bed weight (i.e., less than 1% of bed weight) over the range of conditions studied. The bed sieve analysis data, a selection from which is presented in Table 7.2, showed that

(1) Most of the particle breakdown occurred during the initial few minutes of spouting (less than 10 min), the size reduction beyond this period usually being slight.

(2) The extent of attrition increased sharply with decreasing inlet orifice size. With the two larger orifices (25.4 and 12.7 mm), the original particles did not break down to a very small size, reducing from $-6 +9$ mesh to only $-9 +32$ mesh. A substantial proportion of fine material ($-32 +100$ mesh) was generated only with the 6.3 mm diameter orifice, in which the jet velocity reached nearly 400 m/sec.

(3) For the smaller orifices, particle breakdown with the steeper cone (30° included angle) was significantly more severe than with the larger angle (60°) cone.

(4) Near the minimum spouting velocity, attrition became slightly more severe as the bed depth was increased from 35 to 59 cm, while for a given bed depth, a small enhancement of attrition occurred as the gas flow rate was increased to $1.2U_{ms}$. Both trends are consistent with the predominant effect of orifice velocity brought out by the series of experiments mentioned under (2).

(5) The exact configuration of the air inlet did not affect attrition noticeably. The inlet designs used were both straight and converging pipes, and one arrangement where a pipe protruded for 25.4 mm into the base of the

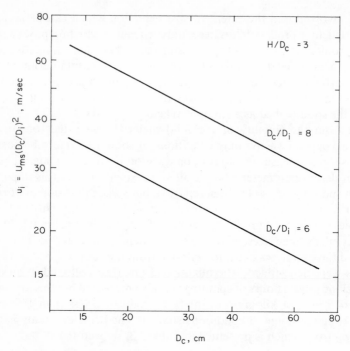

Fig. 7.1. Orifice air velocity for minimum spouting of geometrically similar urea beds, according to Eq. (2.38).

bed (as in Fig. 6.2b). The actual size of the inlet opening in each case was kept the same (15.9 mm).

Thus, from the point of view of particle attrition, by far the most important variable is the diameter of the gas inlet opening, and therefore the velocity at which the gas enters the bed. Reference to Eq. (2.38) shows that this velocity will be smaller in larger columns, for any given values of H/D_c and D_c/D_i. This is illustrated in Fig. 7.1, which shows the dependence of orifice velocity on column diameter for urea, calculated from Eq. (2.38). It is therefore expected that bench scale measurements would show attrition to be more severe than would be encountered in a dimensionally similar larger-scale unit.

8 | Heat Transfer

8.1 BETWEEN FLUID AND PARTICLES

Usually, the fluid–particle interaction in spouted bed processes would involve heat transfer accompanied with either mass transfer or chemical reaction, rather than heat transfer alone. Application to simply heating or cooling of granular solids is, however, possible, and at least one industrial unit for cooling of fertilizers, using a multiple spouted bed, is in operation in Britain [62, see Chapter 11]. Apart from any practical importance, a study of the mechanism by which heat transfer occurs between the bed solids and the spouting fluid is an essential first step toward understanding the more complicated interactions involved in processes such as drying, granulation, and carbonization.

A. Transfer Mechanism

Let us examine the overall mechanism by which heat is transferred from the spouting fluid to the bed solids, assuming for the time being that transfer

131

to individual particles occurs under conditions of external control rather than being controlled by propagation of heat within the particle. Since the solid particles are well mixed, their average temperature in different parts of the bed would be substantially the same. The decline in fluid temperature as the fluid travels up the bed will therefore be governed by the following general equation, which presumes plug flow of fluid:

$$\frac{T_g - T_b}{T_{gi} - T_b} = \exp\left[-\frac{6h_p(1 - \epsilon)z}{U\rho_g c_{Pg} d_p} \right] \tag{8.1}$$

where T_g is the gas (or fluid) temperature at vertical distance z from the gas entrance, T_{gi} the inlet gas temperature, and T_b the bulk solids temperature. Spherical particles have been assumed for simplicity in expressing the particle surface area in terms of d_p and ϵ. Now the spouted bed consists of two distinct regions, the spout and the annulus. In the spout, the average gas velocity U is often one or two orders of magnitude higher than in the annulus, while the volume concentration of particles, $(1 - \epsilon)$, is at most one-fifth of that in the dense phase annulus. As a consequence of the former factor and despite the latter, the heat transfer coefficient h_p in the spout is much higher than in the annulus, but the latter factor has an adverse effect on spout heat transfer rate via particle surface area. The decline of gas temperature in these two regions, according to Eq. (8.1), would therefore be quite different.

Let us apply Eq. (8.1) to each region, using typical spouting conditions. For estimating h_p in the high voidage spout, where the particle Reynolds numbers are generally higher than 1000, an appropriate equation is that of Rowe and Claxton [204],

$$Nu = A + B\,Pr^{1/3}\,Re^{0.55} \tag{8.2}$$

where $A = 2/[1 - (1 - \epsilon)^{1/3}]$ and $B = 2/3\epsilon$, while for the annulus, with its low particle Reynolds numbers (< 100), the packed bed equation of Littman and Sliva [122], which is based on experimental data using air, is more suitable:

$$Nu = 0.42 + 0.35\,Re^{0.8} \tag{8.3}$$

The values of 400 W/m²-°K for the heat transfer coefficient in the spout region estimated from Eq. (8.2), and of 50 for the annulus from Eq. (8.3), are indicative of the orders of magnitude, though considerable variations would obviously occur, depending on the bed material, scale of operation, spouting velocity, etc., as well as on bed level. The general conclusion can nevertheless be reached, by substituting the above values together with typical figures for U and ϵ into Eq. (8.1), that the annulus gas would attain thermal equilibrium with the bed solids within a few centimeters of its entry into the bed, while the distance necessary for the spout gas to achieve equilibrium would be greater by one or two orders of magnitude.

This does not, however, mean that the bulk of the annulus plays no role in the heat transfer process. The function of the annulus is important but indirect and is best elucidated by considering the temperature history of a particle circulating in the bed, assuming a solids heating process. While the time spent by a particle in the spout is an insignificant fraction of that spent in the annulus, the former time is spent under the influence of a much higher heat transfer coefficient and, for a particle entering the spout near the bottom, a higher driving force, as already explained. Particles leaving the spout would therefore be somewhat warmer than, and would discharge their heat on joining, the annular solids. This transfer would occur partly by conduction to the surrounding particles and partly by convection to the gas percolating through the annulus. Thus, the indirect function of the annulus is to serve as a heat sink for the recirculating particles which pick up heat from the spout gas during each cycle. Heat at considerable rates would be transmitted by this mechanism despite the fact that only a small rise in particle temperature (a few degrees) occurs during each pass through the spout, since the solids circulation rates in a spouted bed are very high (several megagrams per hour). A parallel process occurs in fluidized beds where the particles in the region immediately above the gas distributor, which is analogous to the spout region of a spouted bed, attain slightly higher temperatures than the bulk solids and thus act as heat carriers to the rest of the bed.

In continuous operation, the annulus serves an even more important indirect function—it acts as a heat source for the cold feed particles. The time required to bring a feed particle close to the bulk solids temperature is given by the following unsteady state equation:

$$\frac{T_p - T_{p0}}{T_b - T_{p0}} = 1 - \exp\left[\frac{h_p A_p' t}{m_p c_{Pp}}\right] \tag{8.4}$$

where T_p is the particle temperature after residence time t in the annulus, while T_{p0} represents T_p at $t = 0$. The terms A_p', m_p, and c_{Pp} are the particle surface area, mass, and heat capacity, respectively. Using Eq. (8.3) to estimate h_p, we find that the time t from Eq. (8.4) works out to about 1 min for typical conditions. Since practical mean residence times in the slow moving annulus are at least several minutes, the steady state concentration of bed particles having a temperatue significantly below the bulk bed temperature would be small. Therefore, the overall heat transfer rate would not normally be limited by this step of the process.

Let us now examine how the heat transfer process would be modified if transfer between the fluid and individual particles occurs under internal control conditions. Intraparticle temperature gradients are usually ignored in fluidization, but with the large size of particles commonly used in spouting, considerable gradients can build up in certain parts of the bed. When the fluid surrounding a particle is suddenly changed in temperature, the magnitude of the resulting intraparticle temperature differences relative to the temperature differences between the particle surface and fluid is uniquely determined by the Biot number, $Bi_H = h_p r_p / k_p$, provided that the Fourier number, $Fo_H = \alpha t / r_p^2$, which is the dimensionless time variable, exceeds a minimum value of 0.2 [97]. The relative magnitude decreases with decreasing Bi_H, the maximum intraparticle temperature difference becoming less than 5% of the temperature difference between fluid and particle surface at $Bi_H = 0.1$. The absolute magnitude depends also on Fo_H, decreasing with increasing Fo_H. For $Fo_H < 0.2$, the relative magnitude too depends on Fo_H.

The value of Bi_H for millimeter sized particles (of materials like grain, fertilizer, etc.) traveling in the spout is typically of the order of unity, while the maximum value of Fo_H is much less than 0.2. Conditions for the development of significant internal temperature gradients within a spout particle therefore exist. The particle temperature profile is governed by the unsteady state diffusion equation

$$\frac{\partial T}{\partial t} = \frac{\alpha}{r^2} \frac{\partial (r^2 \, \partial T / \partial r)}{\partial r} \tag{8.5}$$

and can be calculated as a function of time for the variable conditions along the spout height by a numerical solution of this equation, if the longitudinal profiles of gas and particle velocities, gas temperature, and spout voidage are known. Since transfer of heat to or from the particle surface occurs by convection, the boundary condition to be satisfied is

$$k_p (\partial T / \partial r)_{r = r_p} = h_p (T_g - T_{r = r_p}) \tag{8.6}$$

where h_p is given by Eq. (8.2).

The outcome of such a calculation carried out by the present authors on a computer for a typical bed is illustrated in Fig. 8.1, with the time coordinate transformed into vertical distance from air inlet and expressed as z/H. The particle considered was assumed to start its upward journey in the spout at $z = 0$ with a uniform temperature which was found by trial and error to satisfy dynamic steady state for the recirculating particle, allowing heat transfer during its stay in the annulus to occur again in accordance with Eq. (8.5), this time using h_p from Eq. (8.3) in the boundary condition, Eq.

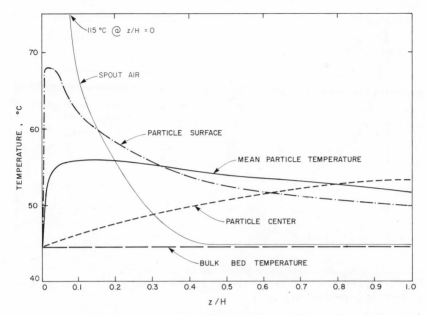

Fig. 8.1. Temperature history of a recirculating particle, including transient intra particle gradients developed during its passage through the spout: $D_c = 30.5$ cm; $H = 61$ cm; $D_i = 5.1$ cm; $r_p = 1.8$ mm; $c_{Ps} = 1.25$ J/gm-°K; $k_p = 0.31$ W/m-°K; $h_{p, spout} = 358$ W/m²-°K for mean conditions over the height of the spout, from Eq. (8.2); $h_{p, annulus} = 85$ W/m²-°K for mean conditions over the height of the annulus, from Eq. (8.3). Note that above a certain level in the spout, the point of maximum temperature within the particle shifts inward, causing the mean particle temperature

$$\overline{T}_p = \frac{1}{\frac{4}{3}\pi r_p^3} \int_{r=r_0}^{r=r} T \cdot 4\pi r^2 \, dr$$

to exceed its surface temperature.

(8.6). The data on air and particle velocities, air temperature, and voidage required for the calculation were all based on experimental observations [137, 138]. The air temperature curve in Fig. 8.1 was in fact observed during drying of wheat [138], where heat transfer is accompanied by mass transfer. Calculations, however, showed that the rate of mass transfer, which was controlled by internal diffusion, was so small in the spout that the particle temperature history therein was not significantly affected by the mass transfer process. The computed results in Fig. 8.1, with h_p evaluated at longitudinal mean conditions for both spout and annulus, show that while a considerable temperature gradient suddenly builds up in the lower-most section of the spout, it gets equalized as the particle proceeds further up, and a small reverse gradient even develops in the upper part of the

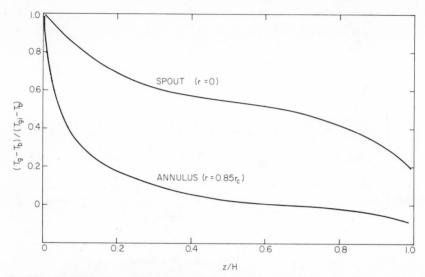

Fig. 8.2. Typical local gas temperature profiles in a bed spouted with hot air, based on measurements by Uemaki and Kugo [235].

spout. The apparent violation of the second law of thermodynamics, whereby the gas in the spout continues to cool in the presence of a hotter particle surface, is explained by the fact that cross flow delivers many colder particles into the spout along the full length of the bed. The temperature of these cross-flowing particles, that is, the bulk solids temperature, is for the present illustration even lower than in the absence of mass transfer in the annulus, and the decline in spout air temperature is therefore also steeper than would otherwise occur (e.g., as in Fig. 8.2).

Thus in deep beds the overall heat transfer process would be only slightly affected by intraparticle temperature gradients, but in shallow beds such gradients could lead to a significant reduction in gas-to-solids heat transfer rate as well as in thermal efficiency.

Turning now to the descent of the recirculating particles, we find that any intraparticle gradients developed in the spout would be effectively relaxed during their slow travel down the annulus, the time required for obtaining a uniform temperature under typical conditions, according to Eq. (8.5), being of the order of 10 sec. As the particle enters the lowest part of the annulus and comes in contact with much hotter gas, internal temperature gradients would again develop. The consequent loss in temperature driving force would cause the decline in gas temperature to be less rapid over this section than in the absence of intraparticle gradients, but with the high heat transfer area in the annulus, thermal equilibrium between gas and solids would be quickly achieved above this section. Therefore, annulus particle temperature gradients should not significantly reduce the thermal efficiency of the process even in relatively shallow beds.

B. Experimental Findings

The mechanism of the heat transfer process postulated in the preceding section is in part supported by the work of Uemaki and Kugo [235], who experimented with 9.2 cm diameter × 12–15 cm deep beds of wheat, coke, glass beads, pumice, alumina catalyst, and silica catalyst (mean $d_p = 1$–4 mm, $\rho_s = 0.93$–2.54 Mg/m^3), using 70°C air as the spouting medium. Steady state conditions were maintained by continuous feed and discharge of solids, and air temperature profiles in the bed were measured using a suction thermocouple. Their typical data, shown in Fig. 8.2, confirm that the decline in gas temperature with increasing bed level occurs more gradually in the spout than in the annulus, and that thermal equilibrium between the gas and solids in the annulus is approached within a few centimeters from the gas inlet. The dip in both curves over the upper part of their slender bed was in all likelihood caused by the effect of fresh feed particles entering from the top, with reduced spout voidage in that region accentuating the dip in the spout curve. Uemaki and Kugo also found that the solids temperature, measured with a bare thermocouple, was almost uniform throughout the bed, as expected. In interpreting their heat transfer data, these workers assumed that the entire heat from the gas is transferred to the solids in the spout region. The temperature driving force for computing heat transfer coefficient was accordingly taken as the integrated average difference between the spout gas temperature and the bulk solids temperature, but the heat transfer surface area, instead of being correspondingly limited to the spout region alone, was based on the total particle surface area in the bed. The values of h_p thus obtained, 3.4–17.0 W/m^2-°K, and the resulting empirical equation for air (Pr = 0.7),

$$\frac{h_p d_p}{k_g} = 0.0005 \left(\frac{d_p U_{ms} \rho_g}{\mu} \right)^{1.46} \left(\frac{U_s}{U_{ms}} \right)^{1.30} \tag{8.7}$$

are therefore rather arbitrary, though the latter does give some indication of the effect on heat transfer of d_p and of U_s, which was varied up to 50% above U_{ms}. Values of h_p for the spout region of the same bed given by Eq. (8.2), using voidages and particle–gas relative velocities estimated from previous experimental data [137] are almost two orders of magnitude greater than those from Eq. (8.7). The ratio between the total number of bed particles and the number of spout particles is of the same order, so that the low values of the coefficient obtained by Uemaki and Kugo are consistent with prediction from Eq. (8.2). All the same, their arbitrary definition of heat transfer coefficient overlooks the indirect function of the annulus in the overall heat transfer process, discussed in the previous section.

The rate at which a cold particle in the annulus receives heat from its surroundings has been measured by Barton and Ratcliffe [10], whose main interest was to predict the temperature history of cold particles continuously fed into a bed of coal maintained under carbonization conditions by spouting with a mixture of air and nitrogen. Once a steady temperature had been reached in the bed, a thermocouple surrounded by a silver bead of the same size as the coal particles (2.5 mm) was introduced at a fixed position within the annular region and the gas-to-particle coefficient was calculated from its transient response by means of Eq. (8.1). Change in bed temperature from one experiment to another was achieved by adjusting the ratio of air to nitrogen in the spouting gas. The steady state temperatures in the annular part of the bed, as measured by a bare thermocouple, were 460, 500, and 550°C in the three experiments carried out. Variation in gas temperature in different parts of the annulus was found to be small (within 20°C, see Fig. 11.17) and an average annulus temperature was taken as T_b in Eq. (8.1). Values of h_p thus obtained were 130–142 W/m²-°K, compared with 102 W/m²-°K calculated from Eq. (8.3). This discrepancy is reasonable considering that the cold stationary test particle was surrounded by hot moving particles, so that in addition to measuring the ordinary packed bed coefficient given by Eq. (8.3), the method used would also register a contribution due to particle convection.

Other experimental work on gas-to-particle heat transfer includes a study of the effect of pulsing the spouting gas flow [56], which is discussed in Chapter 12, and a recent brief report [172] on the comparative performance of fixed, vibrated, fluidized, and spouted beds of 0.6–0.8 mm activated carbon particles in batch experiments over a wide range of hot air flow rates.

C. Design Equations

It has been shown that gas in the annulus would achieve thermal equilibrium with bed solids even in the shallowest bed. If the bed is sufficiently deep, as large-scale beds would normally be, to permit attainment of equilibrium with the surface of the particles in the spout also, then design calculations for gas and solids heating or cooling can be performed without resort to interphase transfer coefficients, as in fluidization. Neglecting intraparticle temperature gradient, and knowing that the solid particles are well mixed, one can assume that $T_{ge} = T_b = T_{se}$. The design for continuous heating or cooling of solids can then be based on the following heat balance equation:

$$W_s c_{Ps}(T_b - T_{si}) = U_s A_c \rho_g c_{Pg}(T_{gi} - T_b) \tag{8.8}$$

For batch heating of cold solids, the heat balance takes the following form:

$$W_b c_{Ps} \, dT_b = U_s A_c \rho_g c_{Pg}(T_{gi} - T_b) \, dt \tag{8.9}$$

which, for initial conditions $t = 0$, $T_b = T_{b0}$, integrates to

$$t = \frac{W_b c_{Ps}}{U_s A_c \rho_g c_{Pg}} \ln \frac{T_{gi} - T_{b0}}{T_{gi} - T_b(t)} \tag{8.10}$$

For the case where thermal equilibrium between exit gas and bed solids is not achieved, it is convenient to define a composite gas-to-particle heat transfer coefficient for the bed as a whole by the equation

$$q = \rho_g U_s A_c c_{Pg}(T_{gi} - T_{ge}) = \bar{h}_p A_p \, \Delta T_{Lm} \tag{8.11}$$

which is equivalent to

$$\ln \frac{T_{gi} - T_b}{T_{ge} - T_b} = \mathrm{St} \frac{A_p}{A_c} = \mathscr{X} \tag{8.11a}$$

where the Stanton number, St, is given by $\bar{h}_p/(c_{Pg} U_s \rho_g)$, or

$$1 - e^{-\mathscr{X}} = \frac{T_{gi} - T_{ge}}{T_{gi} - T_b} = x \tag{8.11b}$$

The term x, which is the fractional temperature approach to equilibrium, can be seen to depend only on St (Re, Pr) and A_p/A_c, that is, only on the bed and particle geometry, gas properties, and operating conditions, but not on the temperature level in the bed (except very slightly via the temperature influence on gas properties). Thus in the absence of thermal equilibrium between exit gas and bed solids, Eqs. (8.8) and (8.10) can be generalized, respectively, to

$$W_s c_{Ps}(T_b - T_{si}) = U_s A_c \rho_g c_{Pg}(T_{gi} - T_b)x \tag{8.8a}$$

and

$$t = \frac{W_b c_{Ps}}{x U_s A_c \rho_g c_{Pg}} \ln \frac{T_{gi} - T_{b0}}{T_{gi} - T_b(t)} \tag{8.10a}$$

For evaluating x from Eqs. (8.11a) and (8.11b), the use of a Stanton number for the bed as a whole is problematical, but a conservative estimate would be obtained by using an average Stanton number for the annulus, with

$\bar{h}_p$ calculated from Eq. (8.3). A rigorous calculation would require writing Eq. (8.11) in terms of separate differential expressions for the spout and annulus regions (on the same lines as in Chapter 10), and would involve the application of Eq. (8.5) if intraparticle temperature gradients are to be taken into account.

8.2 BETWEEN WALL AND BED

Several investigations on heat transfer between column wall and bed have been carried out, since the continuous movement of particles past the column wall provides the opportunity of either supplying supplementary heat to the bed or removing heat from the bed.

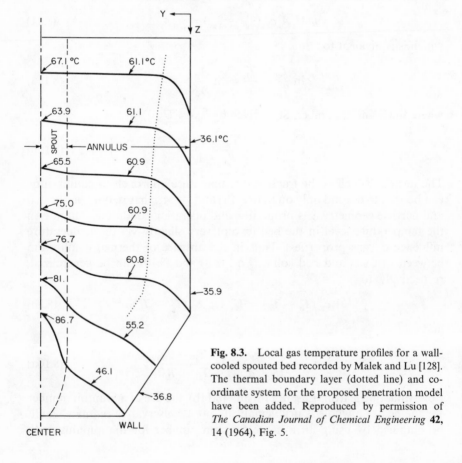

Fig. 8.3. Local gas temperature profiles for a wall-cooled spouted bed recorded by Malek and Lu [128]. The thermal boundary layer (dotted line) and co-ordinate system for the proposed penetration model have been added. Reproduced by permission of *The Canadian Journal of Chemical Engineering* **42**, 14 (1964), Fig. 5.

A. Transfer Mechanism

Transfer of heat from (or to) the wall occurs primarily through convective transport by the downward moving particles in the annulus, while the mechanism by which this heat is further transmitted to the rest of the bed remains the same as in the case of fluid–particle transfer. For spouting with a gas, a characteristic thermal boundary layer no greater than about 1 cm in thickness develops along the wall vertically downward, as illustrated in Fig. 8.3. With the heat transfer zone clearly identifiable, the wall coefficient h_w can be evaluated simply from the general rate equation

$$q = h_w A_w \Delta T \tag{8.12}$$

where q is the heat transfer rate across the column wall, A_w is the area of heat transfer surface, and $\Delta T = (T_w - T_b)$, T_w being the wall temperature and T_b the bed temperature in the annulus measured outside the thermal boundary layer. For the common case of heating or cooling under conditions of high thermal capacity rate of the jacket fluid (condensing vapors or boiling liquids have infinite capacity rate), the wall temperature T_w is independent of bed level; while the annulus temperature T_b varies with bed level only in the lowermost part of the annulus. Therefore, if the heat transfer section surrounds only the upper part of the bed or if it covers a sufficiently small height of the bed, ΔT can be taken as simply the difference between the wall and annulus temperatures. If, however, the heating jacket extends to the lower part, it becomes necessary to use a mean value of ΔT

$$\Delta T = (z_2 - z_1)^{-1} \int_{z_1}^{z_2} (T_w - T_b) \, dz$$

in Eq. (8.12), z_1 and z_2 being the bed levels corresponding to the lower and upper boundaries of the heat transfer section.

B. Experimental Findings

The range of values of h_w evaluated in the foregoing manner by different investigators, alongside the relevant experimental conditions, are listed in Table 8.1.* The first three studies cited in the table are for gas spouting, while the last one concerns liquid spouting. Observed trends in the coefficient of heat transfer h_w are as follows.

---

*Not included in the table are the findings of Gelperin and Fraiman [73], whose study of wall-to-bed heat transfer in conical vessels was restricted to particles of submillimeter size, for which the distinction between fluidization and spouting would be blurred.

TABLE 8.1 Studies of Heat Transfer between Column Wall and Spouted Bed

Bed geometry			Location of heat transfer section	Direction of transfer	System		h_w (W/m^2·°K)	Ref
D_c (cm)	D_i (cm)	H (cm)			Fluid	Solids		
30.5	2.5, 7.6	122	46 cm section at $z = 66$–112 cm	Wall to bed	Air	Wheat, 4 mm	74–91	Klassen and Gishler [100]
15.2, 30.5	2.5–7.6	61, 122				Rice, 3 × 5 mm	68–114	
30.5	1.3, 2.0	122				Sand, 0.6 mm	51–57	
7.6, 15.2	0.5–1.8	25–46	Full bed height	Bed to wall	Air	Wheat, rice, millet, timothy seed, polyethylene, polystyrene, sand; size range of materials 0.8–4 mm	57–136	Malek and Lu [128]
9.2	1.3	25	Full bed height	Wall to bed	Air	Wheat, 4 mm	79–142	Uemaki and Kugo [235]
22.8	1.3	61	Full bed height	Wall to bed	Water	Sand fractions; size range 0.7–1.8 mm	431–568	Ghosh and Osberg [76]

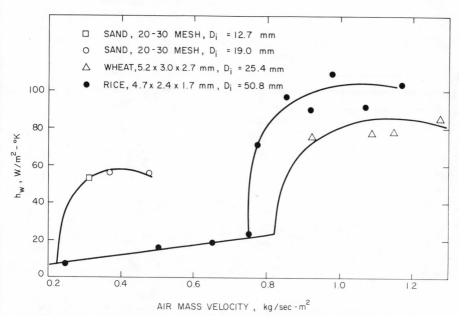

Fig. 8.4. Effect of air flow rate on wall-to-bed heat transfer coefficient [100]. $D_c = 30.5$ cm; $H = 122$ cm. Reproduced by permission of *The Canadian Journal of Chemical Engineering* **36,** 12 (1958), Fig. 6.

1. *Gas–Solid Systems*

(1) The transition from the packed to the spouting state with increasing gas velocity is accompanied by a sharp increase in h_w (see Fig. 8.4), as in fluidization. With further increase in gas velocity beyond the minimum spouting condition, h_w was found either to remain substantially unchanged, as in Fig. 8.4 [100, 128], or to decrease slightly [235].

(2) The coefficient increases with increasing particle diameter, a trend which is the opposite of that observed in fluidization of the same particles (see Fig. 8.5). The behavior for spouting is probably the result of higher particle circulation rates associated with larger particles [38].

(3) For a given material, values of h_w for spouting are lower than for fluidization (Fig. 8.5), the gap between the two becoming narrower with increasing particle size because of the trend noted above.

(4) In two of the investigations [100, 128], neither column diameter nor gas inlet size was found to have a significant effect on h_w, while a small positive effect of inlet diameter ($h_w \propto D_i^{0.2}$) was observed in the third [235].

(5) The average coefficient decreases with increasing bed height; while for a given bed height the localized value increases with increasing distance

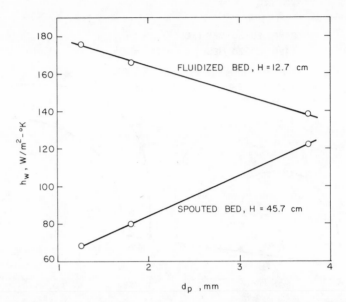

Fig. 8.5. Effect of particle size on wall-to-bed heat transfer coefficient [128]. Material, wheat; $D_c = 15.2$ cm. Reproduced by permission of *The Canadian Journal of Chemical Engineering* **42**, 14 (1964), Fig. 14.

from the gas inlet, as shown in Fig. 8.6. The former trend is explained by an expected increase in the average thickness of the thermal boundary layer with bed height, and the latter by decreasing local thickness and increasing particle velocity at the wall with ascending level in the bed.

2. Liquid–Solid Systems

In the single study of heat transfer to a liquid-spouted bed [76], the thermal boundary layer was found to extend all the way to the spout, as indicated in Fig. 8.7 by the radial profile of bed temperature measured with a bare thermocouple. The arithmetic average of ΔT values at the two ends of the bed was therefore used for computing h_w from Eq. (8.12), with the temperature of inlet water taken as T_b for the lower end and a calculated mixing cup temperature of exit water for the upper end. The main findings of this work, which show the heat transfer behavior of a liquid-spouted bed to be strikingly different from that of a gas-spouted bed, are as follows.

(1) The onset of spouting is accompanied by a sudden decrease in h_w (see Fig. 8.8). This arises from the fact that with water, a large proportion of the fluid (70–80%) passes through the spout; hence, the flow rate through

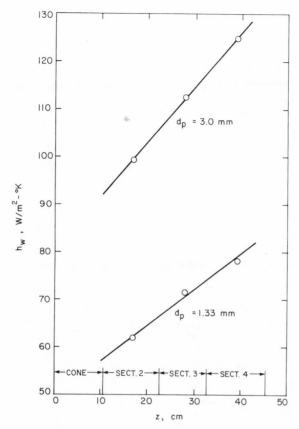

Fig. 8.6. Effect of bed level on local value of wall-to-bed heat transfer coefficient [126]. Material, polyethylene; $D_c = 15.2$ cm; $H = 45.7$ cm. The heating jacket consisted of several separate sections.

the annulus and therefore past the heat transfer surface becomes considerably reduced at the onset of spouting. Although the solids movement brought about by the spouting action would tend to improve the heat transfer coefficient as in the case of gas spouting, the adverse effect of reduced flow through the annulus, being very much greater with water because of its high volumetric heat capacity compared to that of a gas, predominates.

(2) The heat transfer coefficient, after spouting is fully developed, is more or less independent of particle size (Fig. 8.8). However, the decrease in h_w at the onset of spouting mentioned above is less pronounced with fine solids, indicating a more uniform distribution of liquid between the spout and the annulus than with coarse solids. This effect is predictable from the smaller spout diameter which results when the particle size is decreased

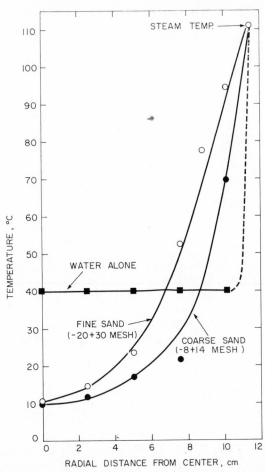

Fig. 8.7. Temperature distribution in a water-spouted bed at $z = 20.3$ cm [76]. $D_c = 22.8$ cm; $H = 61$ cm; $D_i = 1.27$ cm. Water flow rates, gm/sec: Water alone, 454; fine sand bed, 454; coarse sand bed, 756. Reproduced by permission of *The Canadian Journal of Chemical Engineering* **37**, 205 (1959), Fig. 3.

(see Chapter 5) and is consistent with the observed mass transfer behavior of liquid-spouted beds described in the next chapter.

(3) The coefficient for transfer to a water-spouted bed is smaller than to water alone under the same conditions (Fig. 8.8). This is the opposite of the trend which has been reported for heat transfer to water-fluidized beds [35, 195], where the transfer rates were more than doubled by the presence of solids. The slow downward movement of solids without lateral mixing in the spouting bed annulus, according to Ghosh and Osberg [76], appears to suppress the convection currents in the water stream rather than promot-

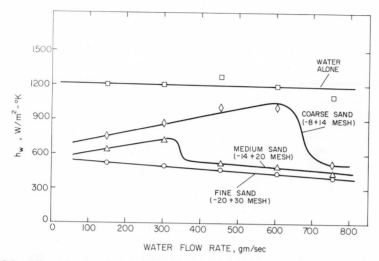

Fig. 8.8. Effect of water flow rate on wall-to-bed heat transfer coefficient [76]. Same system as for Fig. 8.7. Reproduced by permission of *The Canadian Journal of Chemical Engineering* **37**, 205 (1959), Fig. 4.

ing turbulence as in a fluidized bed. They concluded that the heat transfer performance of a water-spouted bed is no better than that of a poorly stirred slurry agitated by a mechanical scraper.

Two different empirical equations for gas spouting, based on dimensional analysis, have been proposed. The equation of Malek and Lu [128], which is based not only on their own data but also on the previous data of Klassen and Gishler [100], is

$$h_w d_p / k_g = 0.54 (d_p/H)^{0.17} (d_p{}^3 \rho_g{}^2 g/\mu^2)^{0.52} (\rho_b c_{Ps}/\rho_g c_{Pg})^{0.45} (\rho_g/\rho_b)^{0.08} \quad (8.13)$$

while the equation of Uemaki and Kugo [235], which is based only on their own data, is

$$\frac{h_w d_p}{k_g} = 13.0 \left(\frac{D_i}{d_p}\right)^{0.2} \left(\frac{d_p U_s \rho_g}{\mu}\right)^{0.10} \left(\frac{d_p{}^3 \rho_g g}{\mu^2}\right)^{0.46} \left(\frac{\rho_s c_{Ps}}{\rho_g c_{Pg}}\right)^{-0.42} (1 - \epsilon) \quad (8.14)$$

The term ϵ in Eq. (8.14) is an overall spouted bed voidage, based on the volume occupied by the bed proper plus the fountain above, and was introduced into the correlation to account for the observed decrease in h_w as U_s was increased beyond U_{ms}. Since the spouted bed annulus, wherein the thermal boundary layer is located, remains a loosely packed bed of substantially constant voidage regardless of the gas flow rate, as discussed in Chapter 5, the use of an overall voidage in the correlation is entirely arbitrary.

TABLE 8.2   Net Dependence of h_w on Individual Variables

Variable	Given by Eq. (8.13) (Malek and Lu [128])	Given by Eq. (8.14) (Uemaki and Kugo [235])
Particle diameter	$d_p^{0.73}$	$d_p^{0.28}$
Bed depth	$H^{-0.17}$	Not varied
Solids heat capacity	$c_{Ps}^{0.45}$	$c_{Ps}^{-0.42}$
Solids density	$\rho_b^{0.37}$	$\rho_s^{-0.42}$
Column diameter	No dependence	Not varied
Gas inlet diameter	No dependence	$D_i^{0.2}$
Gas flow rate	No dependence	$U_s^{0.10}(1 - \epsilon)$

Table 8.2 shows a comparison of Eqs. (8.13) and (8.14) as regards the net effects on h_w of those individual variables which were experimentally varied in each case. It is apparent that serious contradictions arise, pointing to the inadequacy of empirical correlations formulated without reference to the physical mechanism involved.

C. Proposed Theoretical Model

A more rational approach to the problem of wall-to-bed heat transfer, based on the concept of a thermal boundary layer, has been recently proposed [60]. This approach is suggested by the analysis of heat transfer from a cylindrical wall to a moving bed of sand by Brinn *et al.* [31], who simply adopted the theoretical solution [48] for rodlike flow. The annulus of a spouted bed is also a moving bed, but since the thermal boundary layer for gas spouting extends only a small distance from the wall, it is more appropriate here to adopt a two-dimensional penetration model, after Higbie [92]. Neglecting axial conduction relative to transverse conduction, the applicable form of the differential energy equation is

$$v_z \, (\partial T/\partial Z) = \alpha_b(\partial^2 T/\partial Y^2) \qquad (8.15)$$

and from the coordinate system on Fig. 8.3, the boundary conditions are

$$Z = 0, \quad T = T_0 \quad \text{for all } Y$$

$$Y = 0, \quad T = T_w \quad \text{for } Z > 0$$

$$Y = \infty, \quad T = T_0 \quad \text{for all } Z$$

The solution in terms of the surface-mean coefficient h_w over heated length Z is the familiar equation for penetration theory, with the penetration time given by Z/v_z:

$$h_w Z/k_b = 2(Zv_z/\pi\alpha_b)^{1/2} \qquad (8.16)$$

The local coefficient at a particular value of Z is half of that given by Eq. (8.16). The term v_z is the particle velocity at the wall, while α_b is the thermal diffusivity of the loosely packed annular bed given by

$$\alpha_b = k_b/\rho_b c_{Ps} \qquad (8.17)$$

the heat capacity of the bed being essentially that of the solids, since the gas density is negligible relative to that of the particles. Combining Eqs. (8.16) and (8.17) results in the working relationship

$$h_w = 1.129(v_z\rho_b c_{Ps} k_b/Z)^{1/2} \qquad (8.18)$$

To test Eq. (8.18) requires not only a knowledge of ρ_b, c_{Ps}, and Z, which is easily obtained, but also of v_z, the particle velocity at the wall, and of k_b, the effective thermal conductivity of the annular bed. Unfortunately, none of the heat transfer investigators have reported data on v_z, though estimates could be made in a few cases (air–wheat in 15.2 cm and 30.5 cm diameter columns) from the results of other workers [137, 228] on similar beds. The effective conductivity was estimated from literature values of particle conductivity by the method of Kunii and Smith [108], assuming negligible gas flow near the wall in the region of the thermal boundary layer.

Figure 8.9 compares measured values of h_w with those estimated by Eq. (8.18). Unfortunately the only testable data cluster into a small region, but at least the order of magnitude of the theoretical values is correct. In fact, the negative deviation of the experimental results from the theory can be readily explained by postulating a gas gap near the wall approximately 75 microns thick. This hypothetical gap thickness bears a ratio (~ 0.02) to particle diameter intermediate to that postulated by Botterill et al. [26] for beds of flowing solids (~ 0.05) and by Gabor [66] for packed beds (~ 0.01). The method of Schlünder [208] for heat transfer from solid surfaces to moving packed beds should eliminate the need for assuming a hypo-thetical gas gap.

The proposed penetration model predicts several trends in qualitative conformity to those observed, for example, that the local coefficient (which is half the surface-mean value h_w) decreases with increasing Z and hence increases with bed level z, and that h_w is independent of D_i, v_z being sub-

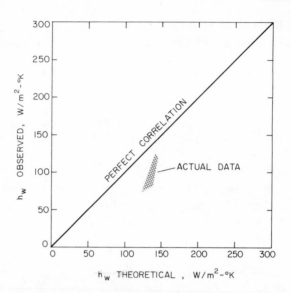

Fig. 8.9. Wall-to-bed heat transfer coefficient: observation versus prediction by penetration model, Eq. (8.18) [60]. Reproduced by permission of *The Canadian Journal of Chemical Engineering* **49**, 467 (1971), Fig. 7.

stantially independent of D_i (see Chapter 4). In its prediction of the positive effect of both c_{P_s} and ρ_b (and hence of ρ_s), theoretical equation (8.18) is far closer to empirical equation (8.13) of Malek and Lu than to Eq. (8.14) of Uemaki and Kugo. Heat transfer data in which the particle velocity at the wall is measured are, however, required in order to give this model a rigorous quantitative test.

8.3 BETWEEN SUBMERGED OBJECT AND BED

Transfer of heat to or from a spouted bed can also be achieved by the use of heating or cooling elements submerged in the bed. For large units this is likely to be a more effective means of heat transfer than the use of a jacket around the column wall, since apart from the geometrical limitation on the heat transfer area per unit volume of bed with the latter arrangement, the relatively long contact time between the particles and the column wall causes the effective driving force to be less favorable than with small submerged heaters installed at suitable locations in the bed.

A. Transfer Mechanism

As for moving and fluidized beds, the two parallel heat transfer mechanisms involved are (1) forced gas convection past the submerged object (a heater, let us say); and (2) particle convection, that is, unsteady state conduction to the particles during their period of contact with the submerged heater, followed by discharge of these particles and their absorbed heat to the bulk of the bed. The relative contribution of the two mechanisms would vary with the location of the heater, depending on local gas velocity, particle velocity, and voidage. In the spout region heat transfer coefficients would be expected to be similar to those in a fluidized bed, while in the annulus they should be comparable (allowing for the differences in size and shape of the heater) to those obtained by Barton and Ratcliffe [10] for a bead immersed in the annulus (discussed in Section 8.1), but somewhat greater than the wall coefficients of Table 8.1. The latter expectation arises not only because of the smaller particle contact times at the heater than at the wall, but also because of the disturbance created by the heater on the gas stream passing it.

B. Experimental Findings

The above reasoning is supported by the extensive experimental results of Zabrodsky and Mikhailik [261], who probed air-spouted beds of silica gel with small electrically heated cylinders, placing the cylinder in both vertical and horizontal positions at different locations in the bed. The heater surfaces were maintained at a constant temperature of 70°C and the current supplied was taken as a measure of the heat transferred. Their main findings are as follows:

(1) Values of the submerged transfer coefficient h_s for the vertical heater at a fixed level in the upper part of the bed were about 30% lower in the annulus than in the spout, the spout values being of the same order as reported by Sarkits [206] for fluidized beds of similar size particles. As shown in Fig. 8.10, the main decrease occurred over a narrow zone, less than 1 cm wide, surrounding the spout.

(2) A slight radial gradient in h_s across the spout itself was also discernible (see Fig. 8.10), values at the spout axis being 4–8% smaller than at the spout–annulus interface. The peak coefficients at this interface could be the result of interaction between the particle velocity and voidage profiles in the spout, both being parabolic in shape, with the lowest values near the

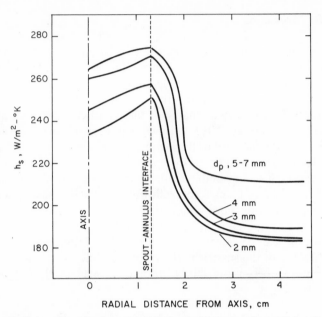

Fig. 8.10. Radial profiles of submerged object-to-bed heat transfer coefficient in upper part of bed measured with a vertically placed cylindrical heater, 0.4 cm diameter × 3.5 cm long [261]: $D_c = 9.4$ cm; $H = 10$ cm; $D_i = 1.5$ cm. Material, silica gel.

interface (see Chapters 4 and 5). Thus the decrease in particle velocity at the spout periphery, which tends to lower h_s, appears to be more than compensated for by the increase in particle concentration, which tends to raise h_s. Zabrodsky and Mikhailik also attributed the peripheral peaks to the ejection of short-circuiting particles from the annulus into the spout.

(3) At a given air flow rate, values of h_s obtained when the vertical heater was placed close to the air inlet with the column empty agreed closely with those determined at the same location with a spouting bed. This result confirms that particle concentration at the inlet is essentially zero (see Chapter 5).

(4) Coefficients measured inside the spout with the horizontal heater were found to decrease sharply with increasing distance from air inlet in the lower part of the spout, but leveled off in the upper part, as shown in Fig. 8.11. Presumably the opposing effects of increasing particle concentration on the one hand (Chapter 5) and decreasing air and particle velocities on the other (Chapters 3 and 4), along the bed height, reach a balance toward the top of the spout.

(5) When the air flow was increased by about 14%, an increase in spout coefficients was registered all along the spout height (see Fig. 8.11). In the

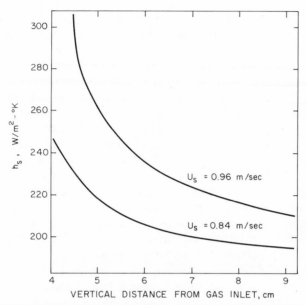

Fig. 8.11. Vertical profiles of submerged object-to-bed heat transfer coefficient in the spout measured with a horizontally placed cylindrical heater, 1.0 cm diameter × 1.7 cm long [261]: $D_c = 9.4$ cm; $H = 10$ cm; $D_i = 1.5$ cm. Material, 2 mm silica gel.

upper part of the bed the fractional increase was about the same as the increase in air flow rate, but in the lower part it was higher.

(6) At the spout–annulus interface in the upper part of the bed, the value of h_s was reported to have increased with increasing air rate until it reached a maximum, and then to have decreased somewhat. It is not clear from the Zabrodsky–Mikhailik paper whether or not the heater was actually moved with each increase in air velocity in order to relocate it at the new position of the spout–annulus boundary. Since the spout diameter is known to increase in direct proportion to the square root of the gas velocity (Chapter 5), failure to relocate the heater with increase in gas velocity would eventually result in its appreciable displacement relative to the boundary, which could contribute to the observed lowering of h_s [see (2)].

(7) Coefficients always increased with particle size, regardless of air flow or heater location (see Fig. 8.10). The authors suggest that the larger particles would cause a greater disturbance of the gas stream in the vicinity of the heater, thereby increasing the contribution to heat transfer by gas convection.

Zabrodsky and Mikhailik made further measurements, using different sizes of heaters and packets of single heaters in the same column as well as in a larger column of rectangular cross section. The latter column was 14 cm × 5.2 cm with a matching conical base of 47° angle, which had the shape of a four-cornered truncated pyramid, the size of the rectangular air inlet opening being 1.2 cm × 5.2 cm. Millet was the only solid material studied, using a bed depth (settled) of 10 cm. Coefficients measured at the spout–annulus interface with the above system were found to be 25–30% lower than the values obtained in the round column at equivalent air flow rates ("same average integral air velocities"). Coefficients in the annulus, on the other hand, were higher in the rectangular column and came somewhat closer to the interface values, the difference being 10–15% with the rectangular column compared to 20–30% with the round column. Heat transfer coefficients determined with a packet heater generally showed good agreement with single heater results, but the insertion of a packet heater in the spout, particularly in its lower part, interfered with the spouting action, causing a reduction in particle velocities in the spout as well as in the annulus, and sometimes asymmetric spouting. The most suitable location for providing submerged heat transfer surface from a practical point of view would therefore appear to be the annular region immediately surrounding the spout, in the upper part of the bed. If a wide heater extending part way into the annulus is required, the use of a column of rectangular rather than circular cross section would offer some advantage.

Other similar investigations in rectangular as well as conical beds but with much finer materials ($d_p < 0.5$ mm) have been reported by Baskakov and co-workers [11, 12]. Particles of the size used in these studies would tend to fluidize rather than spout, and indeed it is reported that in a rectangular bed composed of particles ranging in size from 0.05 to 0.16 mm, a large number of small air bubbles were observed to rise up in the spout, causing a disturbance which spread radially into the annular region [12]. It would appear that although an overall spouting-like pattern of solids movement did exist in these experiments, it was less organized than in true spouting. Nevertheless, profiles of heat transfer coefficient determined with a hot wire probe in the abovementioned bed, as well as profiles measured with a 15 mm diameter copper sphere [12] in conical beds of somewhat larger particles (up to 0.32 mm diameter), showed essentially the same trends as observed by Zabrodsky and Mikhailik with spouted beds of millimeter size particles. Thus, despite the less distinct form of spouting obtainable with fine particles, the internal structure of the beds used by Baskakov and co-workers, as revealed in these heat transfer studies, would appear to confirm the similarity between the structure of their beds and that of a normal spouted bed of coarse particles.

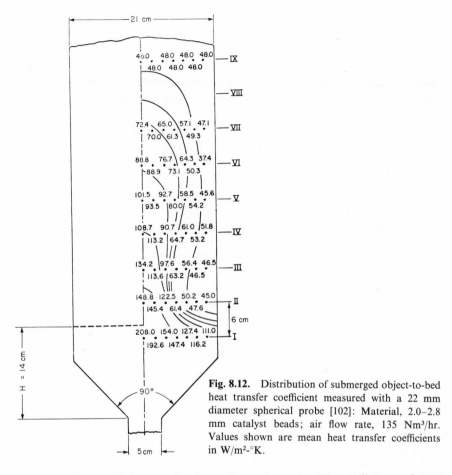

Fig. 8.12. Distribution of submerged object-to-bed heat transfer coefficient measured with a 22 mm diameter spherical probe [102]: Material, 2.0–2.8 mm catalyst beads; air flow rate, 135 Nm³/hr. Values shown are mean heat transfer coefficients in W/m²-°K.

Coefficients of heat transfer from the surface of a 25 mm diameter brass ball have been reported by Klimenko *et al.* [102], with the ball located at various positions not only within the bed but also in the fountain region above the bed surface. The ball probe was so designed that only a small element of its surface served as a heater, so that the distribution of local heat transfer coefficients around the surface of the ball could be determined by rotating the ball around its horizontal axis. The coefficient for a given location in the bed was reported as a mean value over the surface of the ball obtained by integrating the local values measured. The distribution of surface-mean values of h_s obtained is shown in Fig. 8.12, the only radial profile within the bed proper being at level I. Above the bed level the values near the column wall are seen to decrease suddenly by a factor of about 2.5, emphasizing the role of particle convection in the heat transfer. In the axial region the decrease occurs more gradually, presumably because of the presence of a fountain of particles above the bed, coupled with a gradual decay in the spout gas velocity. This decay would increase the velocity in the annular space above the bed, which is therefore seen to temporarily undergo a small rise in h_s.

Distributions of local coefficient around the vertical equator of the spherical probe have also been detailed by Klimenko *et al.* In the axial region of the column, both within and above the bed proper, the values at the front of the sphere were found to be more than twice those at the rear, with a distinct minimum at the 120° position (measured from the front) above the bed, as for an isolated sphere at similar Reynolds numbers [177]. Within the bed the distribution tended to become more uniform as the sphere was moved away from the axis into the annulus, presumably as a result of wake suppression, which is known to occur in a packed bed [243].

9 | *Mass Transfer and Drying*

The discussion of spouted bed mass transfer, unlike that of heat transfer in the previous chapter, is limited to transfer between fluid and particles, surface-to-bed mass transfer being of relatively little practical interest. Mass transfer between fluid and individual bed particles, like heat transfer, can occur under conditions of either external or internal control, but the latter situation is of much greater practical interest since in spouted bed drying applications, the nature of the solid particles and the moisture range involved are often such that the controlling resistance lies within the particle. The emphasis of this chapter, which begins with a brief analysis of external control mass transfer, is therefore on diffusion-controlled drying of solids in a bed spouted with hot gas.

9.1 MASS TRANSFER UNDER EXTERNAL CONTROL CONDITIONS

A. Transfer Mechanism

Mass transfer of a vapor from particles to gas is controlled externally if the vapor originates from drying of the particles within the constant rate

157

period. While the transfer mechanism from an individual particle under these conditions is entirely analogous to heat transfer, no fundamental analogy exists for transfer from the bed as a whole to the spouting gas. This has been pointed out by Beek [18, p. 432] for fluidization and applies equally to spouting. The role of bed particles as heat carriers from hot to colder regions of the bed, discussed in the previous chapter, has no parallel in the mass transfer process, the particles remaining always at the same temperature, namely, the adiabatic saturation temperature of the inlet gas. The question of approach to mass transfer equilibrium between the spouting gas and bed solids can nevertheless be dealt with by analogy with heat transfer, our analysis of the corresponding problem in heat transfer (in Chapter 8) being based on treating the fluid–particle interaction in the spout and the annulus separately, and not on any aspect of the transfer mechanism to the bed as a whole.

The mass transfer analog of Eq. (8.1), with driving force expressed in terms of vapor concentrations of the transferring species (i.e., vapor originating from the solid particles) in the main gas stream, takes the following form:

$$\frac{C^* - C}{C^* - C_i} = \exp\left[-\frac{6K_p(1 - \epsilon)z}{U d_p} \right] \qquad (9.1)$$

K_p being the mass transfer coefficient based on unit surface area of the particles, and C^* the equilibrium concentration at the particle surface. With all the bed particles at the same temperature, namely, the adiabatic saturation temperature, the value of C^* will remain constant throughout the bed. For evaluating K_p in the spout region, the mass transfer version of Eq. (8.2), also due to Rowe and Claxton [204], should be applicable:

$$\text{Sh} = K_p d_p / D_v = A + B \, \text{Sc}^{1/3} \, \text{Re}^{0.55} \qquad (9.2)$$

the definitions of A and B being the same as for Eq. (8.2). For the annulus, the packed bed equation of Thoenes and Kramers [225], with ϵ taken as 0.42, which is valid for both gas and liquid flow over Reynolds numbers between 10 and 1000 [18, p. 443], is adequate for the present purpose even though it may not be the most accurate or general of the numerous correlations to be found in the literature:

$$\text{Sh} = (1.81 \pm 0.36) \, \text{Sc}^{1/3} \, \text{Re}^{1/2} \qquad (9.3)$$

Application of Eq. (9.1), with values of K_p for the air–water vapor system (Sc = 0.6) estimated from Eqs. (9.2) and (9.3) for the spout and annulus regions, respectively, and using the same typical spouting conditions as we had assumed in the heat transfer case, once again leads to the general conclusion that the annulus gas would attain mass transfer equilibrium with the bed solids within a few centimeters of its entry into the bed, and the spout gas within a meter or so.

It should be noted that here, unlike for pure heat transfer, the isothermality of bed solids is not contingent upon the solids being well mixed; rather, it comes about by virtue of each particle being at the adiabatic saturation temperature. The annulus therefore does not play the equivalent of a heat sink role in mass transfer, although in continuous operation it would still serve as a heat source for raising the temperature of fresh feed particles to the prevailing bulk bed temperature in accordance with Eq. (8.4).

B. Experimental Findings

An investigation by Uemaki and Kugo [236], parallel to their heat transfer study discussed in the previous chapter, constitutes the only source of experimental information on particle-to-gas mass transfer under external control conditions. The experiments involved batch drying, over the constant rate period, of silica gel beds, 8 and 10 cm in diameter, spouted with 50°C air. Profiles of air humidity, which is more difficult to measure precisely than temperature, were not determined, so that their results do not lend themselves to directly testing the validity of the transfer mechanism outlined above; but neither does any aspect of the mass transfer behavior observed in this extensive investigation show obvious conflict with the mechanism proposed.

Although Uemaki and Kugo were well aware of the usual difficulties in measuring mass transfer coefficients in dense phase particulate systems, which arise from too rapid attainment of equilibrium between phases, bed depths equal to at least 1 column diameter had to be used in the experiments to obtain the semblance of a stable spouting system. The annulus air at the top of their shallowest beds must have therefore been saturated, while the air emerging from the spout even in their deepest beds, which were less than 0.5 m in height, was found to be unsaturated. The latter observation is consistent with the order-of-magnitude figure of 1 m arrived at in the previous section as the distance required for attainment of equilibrium in the spout region. Actual values of relative humidity of the total outlet gas recorded by these workers for the shallow beds used were 60–75%. Allowing for the contribution of saturated air from the annulus, humidities immediately above the spout must have been somewhat lower, although it appears that the measurement technique used (wet and dry bulb thermometers) was not sensitive enough to detect this difference. Uemaki and Kugo were led to the erroneous generalization that gas particle mass transfer in a spouted bed occurs mainly in the spout region. This error does not, however, invalidate their empirical correlation of the data, which is in any case for an arbitrary mass transfer coefficient, evaluated from bed weight-loss measurements, with the driving force taken as the log mean of terminal concentration differences for the spout region. The particle side of the driving force was assumed to be the saturation concentration of water vapor in air at the measured bulk solids temperature, and the transfer area was taken as the total surface area of all the bed particles. The correlation is as follows:

$$\frac{K_p d_p}{D_v} = 0.00022 \left(\frac{d_p U_s \rho_g}{\mu} \right)^{1.45} \left(\frac{D_c}{H} \right) \tag{9.4}$$

and the experimental conditions for the data supporting it within $\pm 20\%$ are $D_c = 8$ and 10 cm; $H = 1.0\text{--}1.2 D_c$; $D_i = 1.27$ cm; $d_p = 4/6$ to $14/20$ mesh; $U_s = 1.03\text{--}1.7 U_{ms}$; and Re = 20–400.

Numerical values of the Sherwood number (0.03–1.2) for this range of conditions were found by these workers to be at least an order of magnitude smaller than calculated values for fixed and fluidized beds. Here again, as in the case of their heat transfer coefficient, this comparison is not truly indicative of the transfer effectiveness of a spouted bed relative to that of the other systems because of the arbitrariness of their definition of the transfer coefficient. Nevertheless, qualitatively, a loss in mass transfer effectiveness in changing from a packed to a spouting system is only to be expected as a direct consequence of the fact that approach to equilibrium in the spout is much slower than in the annulus, as demonstrated in the previous section. Such a loss was also observed by Mathur and Gishler [139] in some preliminary experiments with a liquid–solid system, involving beds of limestone spouted with dilute hydrochloric acid. Liquid flow rates required in spouting to achieve mass transfer parity with packed states of the same bed were found to be 40% higher with $-8 +14$ mesh limestone and 20% higher with $-14 +20$ mesh material. Concerning liquid versus gas spouting, the contacting effectiveness in the former case is likely to be less favorable in relation to a packed bed, since a liquid, being incompressible, shows a smaller tendency to flare out into the annulus than a gas, as noted by Ghosh and Osberg [76].

To draw any generalized conclusion concerning spouting versus fluidization is more difficult, since it has to be borne in mind that fluid bypassing, which is inherent in spouting, also occurs in gas fluidization, in the form of bubbles, but not usually in liquid fluidization. Thus, while in the liquid spouting work cited above the performance of packed and fluidized beds was found to be similar even at high liquid flow rates, contacting effectiveness in gas fluidization becomes poorer with increasing flow rate because of the bypassing effect mentioned. Consequently, at high flow rates the relative position of gas-fluidized and gas-spouted beds can undergo a change depending upon the extent of bypassing in each system at a specified flow rate. This point is brought out more clearly in the next chapter, in the context of reactor performance.

9.2 MASS TRANSFER UNDER INTERNAL CONTROL CONDITIONS

An interphase mass transfer coefficient loses its relevance if particle-to-fluid mass transfer occurs under such conditions that the transfer rate is controlled by diffusion of the transferring species within a particle, rather than by transfer from particle surface to the surrounding fluid. In practical terms, drying of solids over the falling rate period is controlled by internal diffusion, while theoretically the criterion for internal control is such a high value of mass transfer Biot number, $K_p r_p/\mathscr{D}$, that drying rate becomes dependent only on the corresponding Fourier number, $\mathscr{D}t/r_p^2$, and no longer on the Biot number. Drying of such materials as agricultural products and fertilizer granules, for which the spouted bed has proved to be popular, is often carried out over ranges of moisture content which are well within the falling rate period, estimated values of Biot number under spouted bed drying conditions being of the order of 10^6–10^7 [47]. The question of mass transfer under internal control conditions is, therefore, of considerable practical interest.

The mechanism of the transfer process in this case is inherently more complex than for external control mass transfer, since intraparticle mass transfer is inextricably interlinked with heat transfer—fluid-to-particle as well as intraparticle. While simultaneous transfer of heat is, of course, also associated with external control mass transfer, the heat transfer problem in this case is very much simplified by the absence of both interparticle and intraparticle temperature gradients, the bed solids always being at the adiabatic saturation temperature, regardless of the nature of the solid material. The drying process can therefore be looked upon either as convective heat transfer to isothermal solids, or as convective mass transfer from solids of constant temperature and surface moisture content, as we have preferred to do in the previous section. In contrast, diffusion-controlled drying in a spouted bed is a cyclical process, involving heat transfer to a recirculating particle still in accordance with Eq. (8.5), but with the particle temperature history modified by the simultaneous process of intraparticle mass transfer. The latter is governed by the unsteady state mass diffusion equation [$\equiv$ Eq. (8.5)],

$$\frac{\partial m}{\partial t} = \frac{\mathscr{D}}{r^2} \frac{\partial (r^2 \, \partial m/\partial r)}{\partial r} \tag{9.5}$$

where m represents concentration of, say, moisture at radial position r within a particle and $\mathscr{D}$ the moisture diffusivity. The appropriate boundary condition then becomes the following heat balance equation:

$$k_p \left(\frac{\partial T}{\partial r} \right)_{r=r_p} = h_p (T_g - T_{r=r_p}) - \mathscr{D} \frac{\partial (m\rho_s)}{\partial r} L \tag{9.6}$$

which takes into account the latent heat of vaporization (plus heat of desorption) L, in addition to convective heat transfer. Since $\mathscr{D}$ in Eqs. (9.5) and (9.6) often varies with temperature in accordance with an Arrhenius type relationship, we may write

$$\mathscr{D} = a \exp(-b/RT) \tag{9.7}$$

as a generalization, the values of both a and b being dependent on the nature of the solid material. If $\mathscr{D}$ also happens to be a function of moisture content, as it sometimes is, a further equation to describe this dependence will be necessary.

The full temperature–moisture history, including intraparticle gradients, during one cycle of a recirculating spouted bed particle undergoing diffusion-controlled drying can now be calculated by numerical solution of Eqs. (8.5), (9.5), (9.6), and (9.7), in conjunction with the same hydrodynamic information which was used in the heat transfer calculations leading to Fig. 8.1. The time required for drying a batch of solids to a given mean moisture content,

$$\int_{r=0}^{r=r_p} \frac{4\pi r^2 m \, dr}{\frac{4}{3}\pi r_p^3}$$

can then be rigorously calculated using the particle condition at the completion of one cycle as the initial condition for the next cycle. A treatment of equal rigor for continuous operation is, however, difficult, since here, apart from the residence time distribution of particles in the bed which can be taken into account, the bed at any instant would contain a fraction of particles which are at different stages of drying than the main bed solids.

The cyclical calculations outlined above are too tedious for any practical purpose; and besides, there are often other complicating factors, such as shape and size distribution of particles and their shrinkage, breakage, or agglomeration as drying proceeds. The object of the theoretical analysis is to provide an insight into the complex processes involved in spouted bed drying. For engineering purposes, simplifying assumptions and empirical approximations are obviously necessary, and more practical design procedures, arising out of this necessity, have been proposed. These are presented in the following section.

9.3 DESIGN EQUATIONS FOR SOLIDS DRYING

A. Constant Rate Drying

Equations analogous to those for heating or cooling of solids [Eqs. (8.8)–(8.11)] are required. For deep beds, where the entire gas leaving the bed is saturated with water vapor, $T_{ge} = T_b = T_{as}$, the adiabatic saturation temperature. Design equations, incorporating this condition, can be written either as a heat balance including water evaporation in terms of latent heat of vaporization, or as a water balance.

For a *continuous* drier,

$$U_s A_c \rho_g c_{Pg}(T_{gi} - T_{as}) = W_s c_{Ps}(T_{as} - T_{si}) + W_s L \left(\frac{m_i}{1 + m_i} - \frac{m_e}{1 + m_e} \right) \quad (9.8)$$

where W_s is the feed rate of wet solids; m_i and m_e the inlet and exit solids moisture contents, respectively, expressed as weight fraction of bone dry solids; and L the heat of vaporization of water (plus any heat of desorption). Alternatively,

$$U_s A_c \rho_g (C^* - C_i) = W_s \left(\frac{m_i}{1 + m_i} - \frac{m_e}{1 + m_e} \right) \quad (9.9)$$

the saturation humidity of the gas at T_{as}, C^*, and the inlet gas humidity C_i being in kilograms of H_2O per kilogram of dry gas.

For a *batch* drier,

$$U_s A_c \rho_g c_{Pg}(T_{gi} - T_{as})t = W_b L \left(\frac{m_0}{1 + m_0} - \frac{\overline{m}(t)}{1 + \overline{m}(t)} \right) \quad (9.10)$$

or

$$U_s A_c \rho_g (C^* - C_i)t = W_b \left(\frac{m_0}{1 + m_0} - \frac{\overline{m}(t)}{1 + \overline{m}(t)} \right) \quad (9.11)$$

In the above equations, W_b represents the weight of solid material in the bed and $\overline{m}(t)$ its moisture content after time t.

For the case where spouting gas leaves the bed unsaturated, the following equations can be written by analogy with Eqs. (8.8a), (8.10a), (8.11b), and (8.11a), respectively:

For a *continuous* drier,

$$U_s A_c \rho_g (C^* - C_i) x' = W_s \left(\frac{m_i}{1 + m_i} - \frac{m_e}{1 + m_e} \right) \tag{9.12}$$

while for a *batch* drier,

$$U_s A_c \rho_g (C^* - C_i) t x' = W_b \left(\frac{m_0}{1 + m_0} - \frac{\bar{m}(t)}{1 + \bar{m}(t)} \right) \tag{9.13}$$

where

$$x' = \frac{C_e - C_i}{C^* - C_i} = 1 - e^{-\mathscr{X}'} \tag{9.14}$$

and

$$\mathscr{X}' = \text{St}_M \cdot \frac{A_p}{A_c} = \ln \frac{C^* - C_i}{C^* - C_e} \tag{9.15}$$

The Stanton number, $\text{St}_M = \bar{K}_p / U_s$, is based on a composite particle-to-gas mass transfer coefficient, though here again, as in heat transfer, it would be simpler and safer to evaluate it for the average annulus conditions, using in this case Eq. (9.3).

B. Falling Rate Drying

1. Becker and Sallans

The most generalized design procedure, which in essence is based on the theoretical analysis of Section 9.2, has been proposed by Becker and Sallans [16], and is supported by extensive experimental results for continuous drying of wheat. The main simplifying assumptions made in translating the rigorous theory into working equations, and the restrictions imposed in applying it to drying of a specific material, namely, wheat, are as follows:

(1) Mass transfer in the spout region is negligible relative to that in the annulus, the fraction of total residence time spent by the particles in the spout being insignificant.

(2) The bed solids, being well mixed, are isothermal.

(3) Intraparticle gradients are unimportant and a bed particle is essentially at its mean temperature and mean moisture content.

(4) The moisture diffusion coefficient is independent of moisture content, and varies with temperature in accordance with the following Arrhenius type relationship:

$$\mathscr{D} = 7.66 \times 10^{-3} \exp(-51050/RT) \tag{9.16}$$

with $\mathscr{D}$ expressed in square meters per second, T in degrees Kelvin, and $R = 8.314$ J/(gm-mole)(°K).

(5) The effective surface moisture content for the conditions which commonly prevail in a spouted bed wheat drier (air relative humidity $<40\%$, kernel temperature $>38°C$) has a constant value of 0.103 kg/kg, dry basis.

(6) The wheat particle is treated as a sphere with an effective diameter equal to the product of the equivolume sphere diameter and a shape factor (sphericity) of 0.91.

(7) Change of particle volume as drying progresses is negligible.

It should be noted that with the first three assumptions, the drying rate equations developed, though supported by spouted bed drying data, become applicable generally to drying in any well-mixed isothermal bed.

The general unsteady state diffusion equation was written by Becker and Sallans in the following form to relate moisture content with drying time:

$$\overline{M} = \frac{\overline{m} - m_s}{m_0 - m_s} = \frac{6}{\pi^2} \sum_{n=1}^{\infty} \frac{1}{n^2} \exp\left[-\frac{n^2 \pi^2 \chi^2}{9}\right] \tag{9.17}$$

where $\chi = (A_p'/V_p')(\mathscr{D}t)^{1/2} = (6/\psi' d_v)(\mathscr{D}t)^{1/2}$ for a nonspherical shape (note that for a sphere the term $\chi^2/9$ in Eq. (9.17) reduces to $\mathscr{D}t/r_p^2$, i.e., Fo_M).

In *batch* operation, all the particles are subject to the same residence time, so that Eq. (9.17) would yield the drying time for the whole batch. The bed solids temperature, though uniform at any instant, would increase with time as drying progresses, and so, therefore, would the moisture diffusivity, in accordance with Eq. (9.16). The required differential rate equations, as well as mass and energy balances necessary for the full design calculation, formulated in a manner convenient for integration by computer techniques, have recently been published by Becker and Isaacson [17]. Again, their calculation procedure is applicable generally to any well-stirred wheat drier and not exclusively to a spouted bed drier.

For *continuous* drying, Becker and Sallans combined Eq. (9.17) with the residence time distribution of particles for perfect mixing, derivable from Eq. (4.34), obtaining the following integral result:

$$\overline{\overline{M}} = \frac{\overline{\overline{m}} - m_s}{m_i - m_s} = \int_1^{\cdot 0} \overline{\overline{M}} \, dE\!\left(\frac{t}{\overline{t}}\right)$$

$$= \int_0^{\cdot \infty} \frac{6}{\pi^2} \sum_{n=1}^{\infty} \frac{1}{n^2} \exp\!\left[-\frac{t}{\overline{t}}\left(\frac{n^2\pi^2}{9}\overline{\chi}^2 + 1\right)\right] d\!\left(\frac{t}{\overline{t}}\right) \qquad (9.18)$$

where

$$\overline{\chi} = (A_p'/V_p')(\mathscr{D}\overline{t})^{1/2} \qquad (9.18a)$$

and

$$E(t/\overline{t}) = E(\theta) = -dI(\theta)/d\theta = -d(e^{-\theta})/d\theta = e^{-\theta} = e^{-t/\overline{t}} \qquad (9.18b)$$

The term $\overline{\overline{m}}$ represents the average moisture content of a statistical population of the bed (or product) particles, the moisture content of individual particles being, of course, different because of the residence time distribution associated with good mixing.*

While the integral on the right-hand side of Eq. (9.18) can be evaluated graphically using published solutions, these workers found that for the practical range of values of $\overline{\chi}$ (0.3–1.5), Eq. (9.18) was closely represented by the following much simpler empirical equation:

$$\overline{\overline{M}} = 1 - 1.04\overline{\chi}\exp(-0.44\overline{\chi}) \qquad (9.19)$$

The validity of Eq. (9.19) was demonstrated by showing that diffusion coefficients back-calculated by this equation from spouted bed wheat drying data, covering a wide range of conditions, followed the same dependence on observed bed temperature as given by Eq. (9.16). This agreement gains in significance by the fact that Eq. (9.16) was established independently, on the basis of results from air as well as vacuum drying of fully exposed wheat kernels.

A more direct test of the validity of Becker's theory is represented in Fig. 9.1, which shows a plot of experimental against calculated [by Eq. (9.19)] mean residence times, the latter results being based on values of $\mathscr{D}$ from Eq. (9.16) at observed bulk bed temperatures. Considering the wide range of experimental conditions involved (shown in Table 11.2), the agreement between prediction and experiment is generally quite good (within ±20%), except for two short residence time runs (represented by data

*The question of evenness of drying in a continuous spouting system has recently been examined in detail by Popov et al. [183].

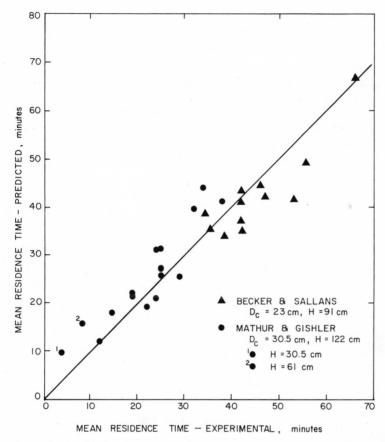

Fig. 9.1. Performance of continuous spouted bed wheat drier as predicted by Eq. (9.19) versus experimental results. Calculation is based on $m_s = 0.103$ gm/gm, dry basis. The experimental conditions covered by the data are shown in Table 11.2, lines 1 and 2 [16] and [138].

points marked 1 and 2 in Fig. 9.1), where the predicted values are much higher. The discrepancy cannot be dismissed lightly since these two runs were conducted [138] as part of a bed depth series with other conditions held constant, and correspond to the shallowest beds used of all the runs represented in Fig. 9.1. The short residence times in these two runs thus arise from the smaller bed holdups rather than from slower feed rates. Possible reasons, involving the structure of a wheat kernel and the role of intraparticle gradients [147], for what amounts to a breakdown of Eq. (9.19) (overprediction of retention times being 150% and 90% for the 30 cm and 61 cm deep beds, respectively) in the case of shallow beds have been discussed elsewhere [60], but without arriving at any conclusive explanation. It appears that some of the simplifications made, in particular, assumptions (2), (3), and (5) (p. 165), cease to be valid for shallow beds.

For deeper beds, the method of Becker and Sallans has been further supported by the experimental results of Quinlan and Ratcliffe [186], which are again for wheat but of the Australian variety. The method has also been successfully applied to the spouted drying of coffee beans [210], the moisture diffusivity of which was found to be about ten times that of wheat. In practice, however, Becker's calculation procedure has not been applied widely to drying of other solid materials because it requires a knowledge of particle moisture diffusivity and its temperature dependence, on which published data are scarce.

A full drier design calculation based on the above method is presented as an appendix in the Becker and Sallans paper.

2. *Kugo, Watanabe, Uemaki, and Shibata*

A more empirical approach for correlating drying rate data, which boils down to dropping all but the first term in the series of Eq. (9.17), has been proposed by Kugo *et al.* [104], on the basis of their experimental work on spouted bed wheat drying, both batch and continuous. The progressive drying of particles in batch operation was described by the following equation:

$$\overline{M} = \frac{\overline{m} - m_s}{m_0 - m_s} = e^{-Kt} \qquad (9.20)$$

and the constant K was empirically correlated with m_0 and T_{gi} (°C), as follows:

$$K = 6.5 \times 10^{-4} m_0 T_{gi} \qquad (9.21)$$

In evaluating K, the value of m_s in Eq. (9.20) was taken as the constant moisture content attained after prolonged drying (equilibrium moisture content). This value varied from one experiment to the other depending on the inlet air temperature used, approximately obeying the following empirical relationship:

$$m_s = 0.16 - 0.0006 T_{gi} \qquad (9.22)$$

with T_{gi} again expressed in degrees centigrade. The range of experimental conditions on which Eqs. (9.21) and (9.22) are based is shown in Table 11.2 and [104].

The final working equation for continuous drying given by Kugo *et al.*, rendered explicit in $\overline{\overline{M}}$, is

$$\overline{\overline{M}} = \frac{\overline{\overline{m}} - m_s}{m_i - m_s} = \int_1^0 \overline{M}\, dE(t) = \frac{1}{1 + K\bar{t}} \qquad (9.23)$$

which is derived by combining Eq. (9.20) with Eq. (9.18b) to allow for residence time distribution of particles assuming perfect solids mixing. Equation (9.23) was tested by Kugo *et al.* against data obtained in continuous runs, which covered the range of their variables as shown in Table 11.2, by comparing the calculated values of $\bar{t}$, using K evaluated from Eqs. (9.21) and (9.22), against experimental values. The calculated results were, however, found to be consistently lower, by 10–15%. Attributing this small discrepancy to the approximate nature of the perfect-mixing assumption contained in Eq. (9.23), Kugo *et al.* devised a correction factor, termed volume efficiency, for use as a multiplier to the calculated values of $\bar{t}$. The numerical value of this factor is variable, being the ratio of hold-back for perfect mixing ($= 1/e$) to experimentally determined hold-back values (see Chapter 4). These values for the spouting conditions corresponding to their drying runs ranged between $0.84/e$ and $0.93/e$, so that the corrected results of $\bar{t}$ from Eq. (9.23) agreed closely with their experimental values.

We have tested the above method against the wider range of conditions presented by the data in Fig. 9.1, and have concluded that it does not have general applicability. The principal weakness of the method appears to be the empiricism involved in selecting T_{gi} rather than T_b as the correlating temperature in Eqs. (9.21) and (9.22). The basic equations of Kugo *et al.*, namely, Eqs. (9.20) and (9.23), are nevertheless as soundly based as Eq. (9.19) of Becker and Sallans.

3. Peterson

It was again on grounds of convenience that Peterson [178], pursuing an earlier approach by Mathur and Gishler [138], directly correlated measured bulk bed-solids temperature with all the independent variables of a continuous spouted bed drier as follows (units in Fig. 9.2):

$$T_b = 0.675 \frac{[1.8(T_{gi} + 32)]^{0.63} d_p^{0.57} D_c^{0.38}}{W_s^{0.24} m_i^{0.29}} - 19 \qquad (9.24)$$

d_p being the geometric mean of three dimensions for nonspherical particles. Corresponding water evaporation rates were correlated with m_i, T_{gi}, and

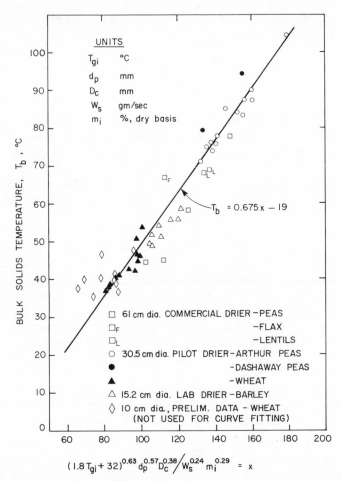

Fig. 9.2. Peterson's correlation of spouted bed drying data. See Table 9.1 for the range of data covered. Reproduced by permission of *The Canadian Journal of Chemical Engineering* **40,** 226 (1962), Fig. 4.

W_s by another empirical equation, but the two equations are not independent of each other, since the evaporation rate for a given run can be deduced from the bed temperature by a simple enthalpy balance calculation, given that $T_b \simeq T_{ge}$. Design calculations can therefore be based on T_b estimated from Eq. (9.24), avoiding the tedious procedure of simultaneously satisfying the rate equation (involving the difficult-to-find moisture diffusivity) and the energy balance equation, required in using Becker's design method.

TABLE 9.1   Range of Experimental Data Supporting Eq. (9.24)

Material dried	d_p (mm)	D_c (cm)	D_i (cm)	H (cm)	m_i (% dry basis)	T_{gi} (°C)	T_b (°C)
Barley	4	15	1.9	46	—	—	49–60
Wheat	3.5	30	6.4	30–122	—	—	36–49
Peas A	6.8	30	5.1	41, 61	—	—	74–104
Peas D	6	30	5.1	41, 61	—	—	81–96
Peas A	6.8	61	8.9, 10.2	216	31	284	78
Yellow peas							
1st pass	7.5	61	8.9, 10.2	216	67	271	58
2nd pass					33	263	74
Seed peas							
1st pass	7.5	61	8.9, 10.2	216	27	125	46
2nd pass					23	124	45
Lentils	4.4	61	8.9, 10.2	216	25	297	69
Flax	2.2	61	8.9, 10.2	216	35	447	67

While Peterson's equation has obviously no semblance of being theoretically based, the range of conditions for which it has proved to be valid (see Table 9.1) with an average deviation of 5% and maximum deviation of 18%, is remarkably wide. The equation should, therefore, have practical value. Quinlan and Ratcliffe [186] have made successful use of this equation for simplifying drier capacity calculations carried out essentially by Becker's method.

4. *Romankov, Rashkovskaya, Kutsakova, Auf, and Frolov*

Several other equations to describe the drying process in a well-mixed isothermal bed, and supported by spouted bed drying data, have been developed by Romankov and associates [201, 105, 106, 5] on the basis of dimensional analysis. Groups involving internal moisture diffusivity were deliberately excluded on the grounds that diffusivity data for solids are not easily available in published literature. The main dimensionless groups used in the correlations are as follows:

(1) Archimedes number (Ar)

$$Ar = \frac{g d_p^{\,3}(\rho_s - \rho_f)\rho_f}{\mu}$$

to account for the external resistance to heat transfer for a particle freely suspended in a gas stream. With gravitational force balanced by frictional drag, Ar for a spherical particle equals $\frac{3}{4}C_D$ Re^2, and is often a convenient substitute for the Reynolds number since it does not contain the velocity term.

(2) Fourier number (Fo)

$$\mathrm{Fo} = \frac{4k_p t}{\rho_s c_{Ps} d_p{}^2}$$

which is a dimensionless time, to account for transmission of heat by conduction into the interior of the particle. (An equation relating moisture content to Fourier number alone, and the geometrical parameter D_c/D_i, has been proposed by other workers [40] for batch drying of peanuts in a spouted bed.)

(3) Kossovich number (Ko)

$$\mathrm{Ko} = \frac{Lm_i}{c_{Ps}(T_{gi} - T_{si})}$$

which represents the ratio of the heat required for evaporating all the moisture in the solids (L being heat of vaporization plus heat of desorption) to the heat required to bring the feed solids up to the maximum possible temperature, namely, T_{gi}.

(4) The ratio k_g/k_p

The Biot number, which together with the Fourier number describes intraparticle temperature gradients, has not been included explicitly but since $\mathrm{Bi} = \frac{1}{2}\mathrm{Nu} \times k_g/k_p$, the group k_g/k_p, together with Ar, covers the effect of the Biot number implicitly, Nu for forced convection being a unique function of Ar for a freely suspended particle in a gas of fixed Prandtl number, e.g., air ($\mathrm{Pr} = 0.7$).

Despite the care in selecting the dimensionless groups, the validity of each empirical equation involving some or even all of the abovementioned groups remains confined to the range of experimental data from which the particular equation is derived. The constant in each equation is bound to be a function of the class of materials studied, but with the neglect of both Bi_M ($= \frac{1}{2}\mathrm{Sh} \cdot D_v/\mathscr{D}$) and Fo_M ($= \mathrm{Fo}_H \cdot \mathscr{D}/\alpha$) or of related groups such as $D_v/\mathscr{D}$ and $\mathscr{D}/\alpha$, it is doubtful that even the exponents on the dimensionless groups correlated would have any general validity. This limitation of the dimensional analysis approach has been recently pointed out by Romankov himself [199, p. 594], who has, however, rightly noted that such equations can still provide a rough assessment of various factors on drying rate.

C. Intermediate Case

For a given solids drier, there is not always a clear-cut answer to the question whether the drying process is occurring in the constant rate or the falling rate period. The controlling resistance to mass transfer may shift during the residence time of a particle in the drier, so that drying may start under external control and later become diffusion influenced and finally diffusion controlled. The situation in a continuous drier is further complicated by the fact that, at any instant, different bed particles may be undergoing drying in different regimes. Under these conditions, the usual method of deciding the kinetic regime from drying rate curves obtained in a batch experiment becomes unreliable, as demonstrated by Romankov and co-workers, in both spouted and fluidized bed experiments [201, p. 269; 199, p. 590]. Because of these factors, formulation of theoretically based drying rate equations for the intermediate case has seldom been successful for a solids drier of any complexity.

The usual approach is apparently either to treat drying in the intermediate regime as if it were diffusion controlled, or to resort to empiricism entirely. In the former case, the moisture diffusivity, e.g., in Eq. (9.18), loses its fundamental significance and assumes the complexion of an arbitrary rate coefficient.

10 | *Vapor Phase Chemical Reaction*

10.1 THE SPOUTED BED AS A CHEMICAL REACTOR

The use of a spouted bed for carrying out vapor phase reactions in the presence of nonreacting solid particles has attracted relatively little attention, despite the fact that it possesses several of the same properties which have been responsible for the widespread application of fluidized beds in this area, namely, intimate gas–solids contact, ease of addition and withdrawal of solids, good agitation and therefore uniformity in bed temperature. Two possible explanations come to mind:

(a) Since heterogeneous gas phase reactions are usually favored by a large surface to volume ratio of the solids, the use of coarse particles and therefore of a spouted bed would not be advantageous for such reactions.

(b) Since a gas phase chemical reactor involves prolonged operation with the same solids, unlike applications involving continuous solids processing, the cumulative attrition of particles due to impact in the high velocity spout would become excessive with most solids.

Although these two considerations would seem to weigh heavily against the use of a spouted bed as a chemical reactor, recent work by Uemaki *et al.*

174

[237, 238] on thermal cracking of petroleum (see Chapter 11) demonstrates that neither one is completely overriding, and indeed conditions in a spouted bed may be more favorable than in other fluid–solid systems for carrying out certain types of reactions.

In order to assess how a spouted bed compares with other competing systems for carrying out a particular reaction, it is necessary to be able to predict the chemical conversion for a spouted bed reactor. With this end in view, Mathur and Lim [144], pursuing an earlier approach by Uemaki [234], have recently proposed a theoretical model of a spouted bed reactor, involving vapor phase reaction in the presence of catalyst or heat carrier particles. The theory proposed for calculating the extent of chemical reaction is quite general, but its usefulness is restricted by incomplete information on certain hydrodynamic features of spouting. The theoretical model is, however, in such a form that this information can be easily incorporated into it as it becomes available. Meanwhile, the available generalizations, limited observations, and even some speculation have been used to demonstrate how the model can be applied to a real situation, and to make certain predictions concerning the effect of the major variables on gas conversion, as well as on the relative performance of spouted, fixed, and fluidized bed systems. This approach leads to some interesting insights into the fluid–solid contacting characteristics of a spouted bed, and incidentally focuses attention on the gaps in the existing knowledge of physical behavior.

10.2 THEORETICAL MODEL FOR PREDICTING CATALYTIC CONVERSION

Since the extent of reaction taking place in the spout and annulus regions would be obviously unequal, these two regions must be treated separately, as depicted in Fig. 10.1. The concentration (moles per cubic meter) of unconverted reactant leaving the reactor can then be obtained by combining the vapors leaving each region. Thus, in terms of the notation of Fig. 10.1,

$$C_e = (U_{aH} A_a C_{aH} + \overline{U}_{sH'} A_s C_{sH'})/U_s A_c \qquad (10.1)$$

while

$$X = 1 - C_e/C_i \qquad (10.2)$$

where X represents the fractional overall conversion which is sought from this analysis. Changes in volumetric flux due to changes in molar flux arising from the chemical reaction(s), as well as due to pressure changes, are ignored, and an isothermal reactor is assumed.

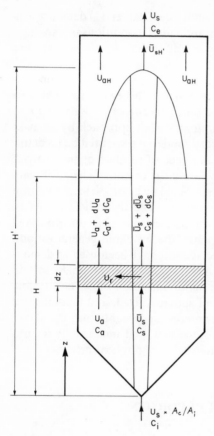

Fig. 10.1. Two-region model of a spouted bed catalytic reactor.

Since U_a, $\overline{U}_s$, C_a, and C_s are all functions of bed level z, it is necessary to write material balance equations over a differential height dz for each region of the bed.

For the spout section dz, under steady state conditions,

Reactant entering section from below per unit time

$\qquad$ = Reactant leaving from above + Reactant cross-flowing into annulus

$\qquad$ + Reactant consumed by catalytic reaction $\hfill$ (10.3)

Consider the simple case of a first order reaction, and porous catalyst particles with no mass transfer and diffusional effects [110, p. 243]. The kinetics of the reaction can then be described by

$$r = -\frac{1}{V_r}\frac{dN}{dt} = K_r C \qquad (10.4)$$

where r is the rate of decomposition of the reactant, moles/(m^3 of solids volume)(sec); V_r is the volume of solids in the reaction zone, m^3; and K_r is the reaction rate constant based on volume of solids, sec^{-1}. Equation (10.3) can now be written in symbols as follows, assuming plug flow of gas:

$$\bar{U}_s A_s C_s = [\bar{U}_s A_s + d(\bar{U}_s A_s)][C_s + dC_s]$$
$$+ \pi D_s \, dz \, U_r C_s + K_r(1 - \epsilon_s)A_s \, dz \, C_s \tag{10.5}$$

where ϵ_s is the spout voidage. From a gas material balance over the section dz,

$$U_r = -\frac{1}{\pi D_s} \frac{d(\bar{U}_s A_s)}{dz} \tag{10.6}$$

Substituting Eq. (10.6) into Eq. (10.5) and rearranging, we get

$$\bar{U}_s(dC_s/dz) + K_r(1 - \epsilon_s)C_s = 0 \tag{10.7}$$

Similarly, for the annulus section dz,

> Reactant entering section from below per unit time
>
> $+$ Reactant entering from spout
>
> $=$ Reactant leaving from above
>
> $+$ Reactant consumed by catalytic reaction $\hspace{2em}$ (10.8)

or in symbols, again assuming plug flow of gas,

$$U_a A_a C_a + U_r \pi D_s \, dz \, C_s = [U_a A_a + d(U_a A_a)][C_a + dC_a]$$
$$+ K_r(1 - \epsilon_a)A_a \, dz \, C_a \tag{10.9}$$

where ϵ_a is the annulus voidage. From a gas material balance,

$$U_r = \frac{1}{\pi D_s} \frac{d(U_a A_a)}{dz} \tag{10.10}$$

Eliminating U_r from Eq. (10.9) with the aid of Eq. (10.10) and rearranging yields

$$U_a \frac{dC_a}{dz} + \frac{1}{A_a} \frac{d(U_a A_a)}{dz}(C_a - C_s) + K_r(1 - \epsilon_a)C_a = 0 \tag{10.11}$$

The effect of gas back-mixing has been neglected in the above analysis. The question of back-mixing of gas, or indeed of any deviations from plug flow behavior in the individual channels, has not yet been fully investigated. If the results of gas tracer experiments such as those in Fig. 3.5 show that significant deviations do in fact occur, one way of taking their effect into account would be by postulating a diffusion type model for the axial spread of gas [110, p. 172] and adding the term $-D\, d^2C_a/dz^2$ to the left-hand side of Eq. (10.11), D being the superficial axial dispersion coefficient.

Equations (10.7) and (10.11) can be solved numerically with the initial boundary condition $C_s = C_a = C_i$ at $z = 0$ to yield $C_{sH'}$ and C_{aH}, provided that the variations in U_a, $\overline{U}_s$, and ϵ_s with respect to z can be quantitatively described. In addition, expressions relating U_s and A_s to the independent variables of the system are also required for substitution in Eq. (10.1) in order to obtain the overall conversion X from Eq. (10.2).

The following relationships were chosen to describe the hydrodynamic features mentioned above.

A. U_a as a Function of z

$$U_a/U_{aH} = 1 - (1 - (z/H))^3 \tag{3.19}$$

$$U_{aH} = U_{mf}[f(H/H_m)] \qquad \text{in accordance with Fig. 3.4}$$

$$H_m/D_c = 0.105(D_c/d_p)^{0.75}(D_c/D_i)^{0.4}\,\lambda^2/\rho_s^{1.2} \tag{6.5}$$

where H_m is in meters and ρ_s in megagrams per cubic meter, λ being a particle shape factor (see nomenclature section of appendix).

B. $\overline{U}_s$ as a Function of z

This can be obtained from Eq. (3.19) and the gas mass balance

$$\overline{U}_s A_s + U_a(A_c - A_s) = U_s A_c \tag{10.12}$$

with the assumption that A_s is independent of z. The conical base of the bed can be taken into account by allowing the cross-sectional area of the annulus, $(A_c - A_s)$, to increase with z until the top of the conical section is reached.

C. ϵ_s as a Function of z

No generalized expression exists and resort has to be made to the limited experimental data in Fig. 5.3. The measurements, however, do not extend beyond $z = H$. For $H < z < H'$, it is arbitrarily assumed that ϵ_s remains at its value for $z = H$.

It is later shown that in a large column (61 cm diameter), spout voidage has only a minor effect on gas conversion. Computations involving a 61 cm diameter reactor have therefore all been carried out assuming an average value of 0.95 over the entire height of the spout, which is close to the longitudinal mean value,

$$\frac{1}{H'} \int_0^{H'} \epsilon_s \, dz$$

for the observed voidage profile represented by curve 2 in Fig. 3.5 ($= 0.97$). For the smaller reactor considered (15.2 cm diameter), the actual observed profile (curve 5, Fig. 3.5) rather than an approximate average value has been used, since here the overall conversion is more sensitive to the extent of reaction in the spout.

D. Minimum Spouting Velocity

$$U_{ms} = \left(\frac{d_p}{D_c}\right)\left(\frac{D_i}{D_c}\right)^{1/3}\left[\frac{2gH(\rho_s - \rho_f)}{\rho_f}\right]^{1/2} \tag{2.38}$$

E. Mean Spout Diameter

$$D_s = \frac{0.118 G^{0.49} D_c^{0.68}}{\rho_b^{0.41}} \tag{5.4}$$

where D_s and D_c are in meters, G in kg/(sec)(m^2), and ρ_b in megagrams per cubic meter. Also, it has been assumed that $H' = 1.25H$ in a 15 cm diameter bed; $H' = 1.50H$ in a 61 cm diameter bed; and the annulus voidage $\epsilon_a = 0.42$.

10.3 PREDICTED REACTOR PERFORMANCE

Equations (10.1), (10.2), (10.7), and (10.11) were solved numerically on a computer, in conjunction with the above hydrodynamic relationships, to explore the effects of the main parameters on gas conversion. The spouting variables used for these computations were selected, as far as possible, to

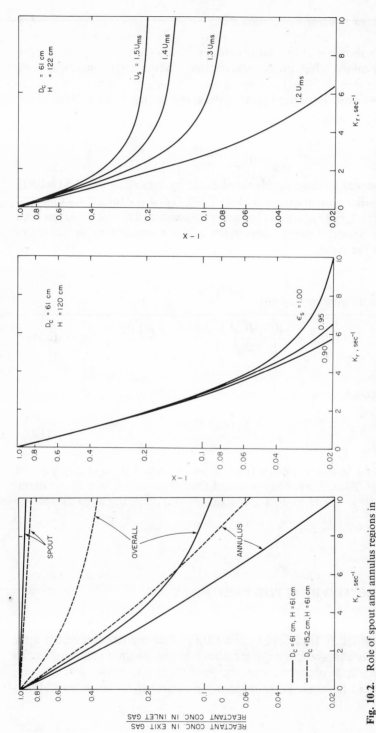

Fig. 10.2. Role of spout and annulus regions in catalytic gas conversion: $d_p = 3.2$ mm; $\rho_s = 1.4$ Mg/m³; $U_{mf} = 0.82$ m/sec; $U_s = 1.2 U_{ms}$; ϵ_s, experimental data (Fig. 5.3, curve 5) for the small reactor, assumed value of 0.95 for the large reactor.

Fig. 10.3. Effect of spout voidage on gas conversion: $d_p = 3.2$ mm; $\rho_s = 1.4$ Mg/m³; $U_{mf} = 0.82$ m/sec; $U_s = 1.2U_{ms}$.

Fig. 10.4. Effect of spouting velocity on gas conversion: $d_p = 3.2$ mm; $\rho_s = 1.4$ Mg/m³; $U_{mf} = 0.82$ m/sec; $\epsilon_s = 0.95$ (assumed).

180

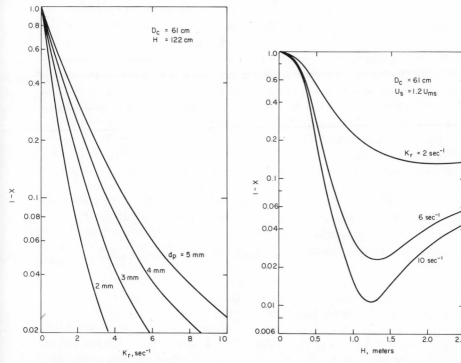

Fig. 10.5. Effect of catalyst particle size on gas conversion: $\rho_s = 1.4$ Mg/m³; $U_s = 1.2U_{ms}$; $\epsilon_s = 0.95$ (assumed).

Fig. 10.6. Effect of bed depth on gas conversion: $d_p = 3.2$ mm; $U_s = 1.2U_{ms}$; $\epsilon_s = 0.95$ (assumed).

match the conditions for which the maximum hydrodynamic data are available, e.g., 15 cm and 61 cm diameter reactors with a conical base of 60° included angle, catalyst of the same size and density as wheat, and vapors of the same properties as room temperature air. The choice of the range of K_r values was dictated by the desire to bring out clearly various trends of the effect of spouting conditions on gas conversion, though the values used are in line with those reported for some industrially useful reactions [110, p. 228]. The calculated results are presented graphically in Figs. 10.2–10.6. With the exception of Fig. 10.6, the results hold no surprises, but they do bring to light some interesting features of a spouted bed catalytic reactor, which are discussed below.

(a) For the same bed depth, the performance of a laboratory size reactor is much worse than of a large unit (Fig. 10.2). The extent of conversion of the spout gas in the two cases is not widely different, but in the larger unit where the spout occupies a much smaller fraction of the bed cross section, the spout gas plays a lesser role in determining the overall conversion. This fraction for the systems of Fig. 10.2, using Eq. (5.4), works out to 1/11 for the smaller and 1/60 for the larger bed. Also, conversion in the annulus region of the larger bed is more complete, since the residence time of the gas is longer than in the smaller bed, as implied in the relationships of Eq. (6.5), Fig. 3.4, and Eq. (3.19). According to Eq. (6.5), H_m for large D_c is higher and therefore, for the same H, Fig. 3.4 gives a lower value of U_{aH}/U_{mf} for the larger bed. Since U_{mf} is independent of column geometry, Eq. (3.19) then shows that U_a at any value of z/H would be lower for the larger bed, causing the residence time of gas in the annulus to be longer.

(b) While the importance of spout voidage decreases with increasing reactor size because of the smaller contribution of spout gas to overall conversion as mentioned earlier, it increases with increasing reaction rate. The latter trend is brought out by the results in Fig. 10.3, which however show that over the likely range of spout voidages in a large bed, the effect of voidage on overall conversion remains relatively small even for the fastest reactions considered. The spout voidage values shown in Fig. 10.3 refer to average values over the height of the spout.

(c) The overall conversion becomes poorer as the operating gas velocity is increased above U_{ms} (Fig. 10.4). This is consistent with the physical picture of the excess gas, $U_s - U_{ms}$, passing through the spout which follows from the correlation in Fig. 3.4, whereby U_{aH} is independent of U_s. The additional effect of spout diameter variation with gas flow rate, in accordance with Eq. (5.4), is also reflected in the results represented by Fig. 10.4.

(d) The performance of the reactor, with U_s maintained at $1.2U_{ms}$, is improved by decreasing the catalyst particle size (Fig. 10.5). This trend is based on K_r being independent of particle size and is therefore simply indicative of a more favorable gas flow pattern with smaller particles. This arises mainly from the effect of d_p on U_{ms} [Eq. (2.38)], and partly through the combined effects of d_p on H_m [Eq. (6.5)], D_s [Eq. (5.4)], and U_{mf}. Values of U_{mf} for particles of different size, required when using Fig. 3.4, were estimated by Eq. (2.36).

(e) The results in Fig. 10.6 show that the performance of the reactor improves with increasing bed depth, but only up to a depth of 1.25 m. For a slow reaction, the use of deeper beds does not bring about any further significant improvement, while for a fast reaction, the conversion actually becomes poorer as the bed depth increases beyond 1.25 m. This unexpected behavior can be traced back to the fact that the change in $\overline{U}_{sH'}$ in Eq. (10.1) with respect to bed depth is caused by the combined effect of H on U_{ms}

and on U_{aH}. The velocity U_{ms}, according to Eq. (2.38), is directly proportional to $H^{1/2}$ but the dependence of U_{aH} on H, as given by Fig. 3.4, becomes less pronounced for increasing values of H. It is largely the net effect of these two separate factors which causes $\overline{U}_{sH'}$ to first decrease and then increase with H, and this in turn gives rise to an optimum value of bed depth in Fig. 10.6. It should be recalled, however, that the evidence supporting the correlation in Fig. 3.4 is not too strong.

10.4 COMPARISON WITH FIXED AND FLUIDIZED BEDS

In order to achieve good solids agitation and consequent uniformity of bed temperature, the size of catalyst particles in a fluidized bed reactor would have to be much smaller than in a spouted bed reactor. The relative performance of these two systems would therefore depend not only on their respective hydrodynamic features but also on the effect of catalyst particle size on reaction kinetics. The assumed independence of reaction rate from particle size, implied in Eq. (10.4), applies to the limiting case of idealized porous catalyst particles with no mass transfer or diffusional effects, where the external surface of particles is negligible in comparison with the internal surface. At the other extreme, one should also consider the case of non-porous catalyst with the reaction occurring on the external surface only. The rate equation for the latter case becomes [110, p. 243]

$$r = -\frac{1}{V_r}\frac{dN}{dt} = \frac{6}{d_p}K_r'C \tag{10.4a}$$

where K_r' is the reaction rate constant based on the external surface of equivolume spheres, in centimeters per second.

A comparison of the performance of spouted bed and fluidized bed reactors of equal size has been made for the two limiting cases mentioned above, with the size of catalyst particles for the fluidized system kept sufficiently small for good fluidization ($d_p = 0.3$ mm). Gas conversion for the fluidized reactor, operating at the same gas throughput rate as the spouted bed, was calculated using the bubbling bed model of Kunii and Levenspiel [110], with the effective gas diffusivity taken as 0.5 cm^2/sec and the effective bubble diameter, which is an adjustable parameter of the model, assumed as 10 cm. Both these values are arbitrary, though of reasonable magnitude, and the predicted performance of the fluidized bed is to that extent speculative.

Table 10.1 shows calculated values of the unconverted fraction of reactant for the fluidized bed and for the spouted bed, for a slow reaction and a fast

TABLE 10.1 Relative Performance of Spouted, Fluidized, and Fixed Bed Reactors for the Same Gas Throughput[a]

	Spouted bed	Fluidized bed	Fixed bed
Catalyst, d_p (mm)	3.2	0.3	0.3
ρ_s (Mg/m^3)	1.4	1.4	1.4
Operating gas velocity (m/sec)	0.56	0.56	0.56
	($=1.2 U_{ms}$)	($=17 U_{mf}$)	
	Fraction unconverted reactant, $1 - X$		
Idealized porous catalyst			
$K_r = 1.0$ sec^{-1}	0.40	0.41	0.28
$K_r = 10.0$ sec^{-1}	0.01	0.06	3.1×10^{-6}
Nonporous catalyst			
$K_r' = 0.004$ cm/sec	0.93	0.47	0.36
$K_r' = 0.04$ cm/sec	0.50	0.07	3.9×10^{-5}

[a] $D_c = 61$ cm, $H = 122$ cm.

reaction. Corresponding values for fixed bed operation assuming plug flow of gas, calculated by the equations [110, p. 244]

$$1 - X = \exp\left[-\frac{(1 - \epsilon)HK_r}{U} \right] \tag{10.13}$$

and

$$1 - X = \exp\left[-\frac{6(1 - \epsilon)HK_r'}{d_p U} \right] \tag{10.14}$$

are also included in the table, for $\epsilon = 0.42$ and $U = U_s = U_f$.

The results in Table 10.1 illustrate the fact that the spouted bed, in common with a fluidized bed, is always a less efficient system for gas conversion than a hypothetically isothermal fixed bed, since agitation of solids is achieved at the expense of inefficient conversion of gas passing through the spout in a spouted bed and as bubbles in a fluidized bed.

With the idealized porous catalyst, where the reaction rate is independent of particle size, the spouted bed is seen to give higher conversions than the fluidized bed, the difference being substantial for the faster reaction considered. This behavior follows from the fact that in order to match the high gas flow rates permitted by the use of coarse particles in a spouted bed, an equivalent fluidized bed with its finer particles must be operated at high multiples of U_{mf}. With such operation, gas conversion becomes poor because of excessive bypassing of gas in the form of bubbles, and consequently spouted bed operation becomes the more attractive under the specific conditions chosen to bring out this effect in Table 10.1. For the case of nonporous catalyst, values of K_r' chosen for the comparison are again arbitrary, but are consistent with the first catalytic case inasmuch as lower conversions are to be expected in beds of nonporous catalyst having the same particle size as porous catalyst, because of the smaller total surface area of the former. Here, the results in Table 10.1 show the spouted bed to be at a serious disadvantage in comparison with the fluidized bed, since the adverse effect of smaller external surface area of large spouted bed particles overshadows the benefit arising from improved gas–particle contact mentioned above.

Hence, the possibility of achieving better conversion by using a spouted bed as a catalytic reactor instead of a fluidized bed would seem to be confined to relatively fast reactions for which the reaction rate is independent of, or perhaps only weakly dependent on, catalyst particle size.

10.5 EXPERIMENTAL SUPPORT

Unfortunately, none of the experimental data from spouted bed studies of gas conversion are suitable for testing the validity of the theoretical model proposed. The kinetics of the petroleum cracking process studied by Uemaki *et al.* [237, 238] are far too complex, since many competing reactions are involved. Realizing this complication, Uemaki [234], who was the first to attempt the modeling of a spouted bed chemical reactor, experimented with the water gas reaction to obtain data suitable for testing his model. This reaction, however, proved to be too slow even at temperatures up to 830°C, the rate constant at this temperature being of the order of 0.1 sec^{-1} with the 1.52 mm coke particles used. The fraction of unconverted steam in his 10 cm diameter reactor over the entire range of operating conditions remained above 0.97, which, in the light of the computed result in Fig. 10.2, is not surprising. The decomposition of ozone, with its simple kinetics and moderately fast rate, is a good reaction for studies of this type and has been traditionally used in fluidization. Volpicelli [249] did carry out this reaction in a spouted bed (primarily to study the effect of gas pulsations—see Chapter 12), but even these data do not lend themselves to testing of the model, since the hydrodynamics of the small two-dimensional reactor used by Volpicelli (186 mm × 8 mm) would not be correctly described by the generalizations based on the behavior of three-dimensional beds, incorporated into the model.

In conclusion, it should be noted that a model of the type proposed by Mathur and Lim [144] should provide a rational basis for analyzing not only chemical reaction but also heat and mass transfer processes in spouted beds. Their two-region model, despite its unproven status, has been presented here in full with the hope that it might stimulate further work in all three areas.

11 | *Applications*

11.1 INTRODUCTION

The operations for which the use of a spouted bed has already attracted attention are remarkably varied. The main features of the various applications are summarized in Table 11.1, which also shows the stage of development in each case. For drying, heating, and cooling of granular solids, and for gas cleaning, the chief attraction of a spouted bed is the same as that of a fluidized bed, namely, good solids agitation combined with effective gas–solids contact. For these operations, as also for solids blending, spouting serves the same purpose for coarse particles as fluidization does for fine materials. In coating and granulation, the main advantage arises from the regular cyclic motion of solids which permits deposition of successive layers on the particles, allowing enough residence time in the annulus between depositions for each layer to dry before the next one is deposited in the spout; while the attrition caused by interparticle collisions in the spout pays a key role in drying of suspensions and solutions onto inert particles, as well as in comminution, coal carbonization, shale pyrolysis, and iron ore reduction. The application of spouting to thermal cracking of petroleum, where the short residence time of the vapors in the bed is a

TABLE 11.1 Summary of Reported Spouted Bed Applications

Application	Main features	Stage of development
Drying of granular materials (references in Table 11.2)	Particularly suitable for heat-sensitive materials such as agricultural products and polymers. Agitation of solids permits the use of high temperature air and therefore rapid drying, without the risk of thermal damage. Also suitable for sticky solids, since agglomerates disintegrate in the high velocity spout.	In industrial use
Granulation [19, 20, 167, 201]	Melt or solution is atomized into the bed, which is spouted by hot gas. Initial bed consists of product granule nuclei, which build up by a mechanism of layer-by-layer growth as they cycle in the bed. A well-rounded granule of uniform structure is thereby produced.	In industrial use
Drying of suspensions and solutions [193, 201]	Same principle as above except that bed consists of inert particles, e.g., glass beads. Solution is atomized into the lower region, coats the particles, the coating becoming progressively more fragile as it dries until it is knocked off by interparticle collisions. Fine product is collected overhead.	In industrial use
Reaction–granulation [255, 229, 230, 41]	Process combines fertilizer granulation with neutralization in the same spouted bed unit. Acid spray and ammonia vapors are introduced with the hot air jet. Bed consists of product granules.	Preliminary work on bench scale
Tablet coating [216, 217, 91, 65]	Batch operation. Coating solution introduced with hot air as in granulation. Compared to conventional coating pans, spouted bed gave more uniform coating, better batch-to-batch uniformity, shorter batch time, and lower overall cost.	In industrial use
Coating of nuclear fuel particles [14, 25, 24, 1, 3]	The bed of fuel particles is spouted with hydrocarbon vapors plus helium, and is maintained at a high temperature ($\sim 1500°C$) by heat input from a surrounding furnace. Pyrolytic deposition of carbon occurs, giving a uniform and tough coating.	Laboratory scale investigations. Effect of process conditions on microstructure of coating has been studied in detail.

Application	Description	Status
Gas cleaning [165]	Spouted bed used for regenerating spent adsorbent (activated charcoal granules) in conjunction with a fluidized bed for vapor adsorption in a gas cleaning system. Solids are fed into the base of the bed by entrainment with spouting gas, and leave by overflow from the top. The required vertical transport of solids is thus achieved at one-tenth the gas velocity for pneumatic transport, in addition to regeneration.	Not known
Preheating of coal [130]	Coal of about 6 mm size heated to 250°C in continuous operation, as pretreatment before coking in coke ovens. Promising outcome, multistage operation visualized for large scale.	Preliminary study in a 15 cm diameter column
Cooling of fertilizers [62]	Double-deck installation with multiple spouts for cooling up to 30,000 kg/hr of fertilizer from 120°C to 40°C.	In industrial use
Solids blending [27, 95]	Rapid and effective blending of granular solids (polymer chips) achieved at lower power cost than in mechanical blenders. Economics of spouted bed blending is particularly favorable for large batch sizes.	In industrial use
Comminution [143]	Particulate solids are finely ground by adding to the bed a proportion of hard and heavy particles of another material (e.g., glass beads) to serve as the grinding medium. Ground product is elutriated by the spouting air and is collected overhead. Advantage over conventional mills visualized for duties requiring grinding simultaneously with intimate gas–solids contact, e.g., for drying, cooling, heating, or reaction.	Preliminary experiments in a 15 cm diameter column
Low-temperature coal carbonization [33, 134, 190, 9, 10]	Use of coarse particles (2.5 mm) together with the violent agitation in the spout region allowed continuous operation without the problem of agglomeration. Process worked well with a variety of Australian coals at carbonization temperatures of 450–650°C.	Detailed investigation in a 15 cm diameter carbonizer
Shale pyrolysis [22, 23]	Continuous operation with coarse shale of up to 6 mm size, temperatures of 510–730°C. Attrition of particles in the spout was beneficial, since the outer surface of a particle became fragile on loss of organic matter and was broken off, exposing fresh surface for retorting. Fine spent shale collected in an overhead cyclone.	Bench scale operation showed spouted bed process to be feasible

TABLE 11.1 (continued)

Application	Main features	Stage of development
Iron ore reduction [7, 245–247]	Good solids agitation, absence of fines in the reaction zone, and breakdown of agglomerates in the spout, all together, enabled reduction to be carried out at temperatures up to 1000°C without the solids becoming sticky. More rapid and complete reduction of coarse ore, as well as of ore–fuel pellets, to sponge iron was consequently achieved than occurs in fixed or fluidized beds.	Effect of process variables has been thoroughly studied, though on a very small scale
Cement clinker production [90]	Decarbonated cement granules of 1–3 mm diameter were spouted at 1350–1550°C. The gas inlet tube was used as clinker product outlet, the terminal velocity of the granules increasing sufficiently during the process for clean-cut separation of the product, but too rapidly for complete conversion of the feed.	Preliminary bench scale work, to be followed by experiments at 1550–1750°C
Charcoal activation [52]	Spouted bed process developed for producing coarse granular activated carbon using steam and flue gas at 1000–1500°C as spouting medium. Batch industrial unit in operation since 1970, capable of activating a charge of 150–200 kg in 80–90 min.	In industrial use
Thermal cracking of petroleum (crude oil, heavy oil, and naphtha) [237, 238]	Bed of 1–4 mm size heat carrier particles was spouted with steam plus petroleum vapors to produce ethylene and propylene at 550–860°C. Use of coarse solids allowed operation at higher gas velocities, hence with shorter contact times, than in fluidized or packed beds. Ethylene yield, being favored by a short contact time, was consequently higher. Heat economy was improved by the use of a spouted bed combustion chamber in conjunction with the cracker.	Optimum process conditions have been determined on a bench scale.

critical requirement, stands by itself, relying on the use of coarse heat carrier particles which permit high gas flow rates to be used.

An account of the work carried out on each of the above applications, under the broad categories of physical operations and chemical processes, follows. A few suggestions for new applications are also included.

PHYSICAL OPERATIONS

11.2 DIFFUSIONAL

The processes grouped together under this heading all involve mass transfer accompanied by heat transfer. The diffusing substance in most cases is water and its direction of transfer from the solid phase to the spouting gas; in other words, drying occurs. Two of the processes described, however, desorption and thermochemical deposition, involve movement of nonaqueous vapors, with diffusion occurring in the opposite direction in the latter case. Other practical applications involving gas to solid diffusion, e.g., vapor adsorption and gas dehumidification, are proposed at the end of this section.

A. Drying of Granular Solids

By far the most popular application of spouted beds has been drying of coarse, heat-sensitive granular materials which include a range of agricultural products, wood chips, various polymeric materials, ammonium nitrate, and manganese chloride. The available data for both experimental and industrial driers are summarized in Table 11.2. Although the range of moistures and drying conditions varies widely, it is seen that there is always a wide temperature gap between the hot air and the bed. This feature of spouted bed drying represents its main advantage over conventional non-agitated driers, in which the smaller temperature gap necessitates that the air temperature be kept much lower in order to avoid thermal damage to the particles.

The pilot wheat drier used by Mathur and Gishler [138], which may be regarded as typical of a continuous granular-solids drying system, is shown in Fig. 11.1. Wet solids enter the upper part of the annulus and become

TABLE 11.2 Available Data on Spouted Bed Drying of Granular Solids

Material dried	d_p (mm)	ρ_s (Mg/m³)	Bed geometry D_c (cm)	Bed geometry H (cm)	Moisture content (% dry basis) Feed	Moisture content (% dry basis) Product	Feed rate (kg/hr)	Air rate (kg/hr)	Air temp (°C)	Bed temp (°C)	Ref
Bench or pilot scale											
Wheat	4.0	1.4	30	30–122	20–36	17–29	110–270	242–278	100–177	38–54	[138]
Wheat	3.6	1.5	23	91	25	15–19	28–43	143–156	97–166	45–75	[16]
Wheat	4.1	1.4	15	9–21	20–44	4–12	Batch	35–53	60–180	–	[104]
Wheat	4.1	1.4	15	14	21–30	14–18	1.8–3.6	35	70–145	–	[104]
Coffee beans	11 × 20	0.9	15	61	45	12	6.5	38	160	68	[210]
Peanuts (whole)	11 × 23	0.3	30–61	–	30	15	Batch	–	–	–	[40]
Wood chips	<16	–	30	38–46	150–190	49–58	36–59	360	400–600	75–88	[42]
Styrene polymers	0.6–1.2	1.0	Conical θ = 24°–50°	8–25	1–30	0.2–0.3	0.5–5	108–192	80–160	–	[106]
Polyvinyl chloride	0.4	1.0			–	–					
Polyvinyl formal	1.8	1.2			–	–					
Colcothar	0.7	3.5	22	17–28	3–290	–	0.2–3.5	3–85	70–120	–	[5]
SG-1 copolymer	2.2	1.3				–					
MnCl₂·4H₂O (dehydration)	2–5	–	10	<43	–	–	0.5–5	12–24	350–500	220–230	[101]
Activated carbon	0.6–0.8	0.4	6	15	12	0.2	0.19	1.7	110	–	
Superphosphate	3–6	0.8	17.6	45	12	4–5	48	–	155–160	–	[166]
Gelatin (two-step process)	>5	–	17.6	8	59	30	0.53	157	28	–	
				25	30	15	0.91	–	50	–	
Industrial scale											
Peas	6.7–7.5	1.4	61	214	23–68	21–47	1000–1700	2600–3100	124–284	45–78	
Lentils	4.4	–	61	214	25	17	1160	1470	297	69	[178]
Flax	2.2	1.1	61	214	35	14	890	1600	448	67	
Ammonium nitrate	−10 +35 mesh	1.5	61	117	7–8	<1	100–125	410	155–165	65–75	[96]
Polymer granules	–	–	122	~244	Other information not released						[53]

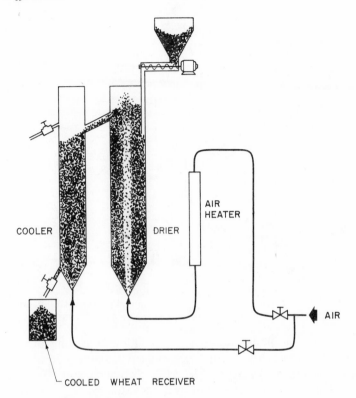

Fig. 11.1. Pilot wheat drier of Mathur and Gishler [138].

rapidly intermixed with the bed material, which discharges continuously through an overflow pipe so that the desired bed level is maintained. The overflow is located diametrically opposite the feed point in order to minimize short circuiting. Hot solids leaving the drier are brought in contact with cold air in a second column to cool them down to a safe storage temperature. Some additional drying also occurs in the cooler by virtue of the sensible heat available in the particles. Since uniformity of bed temperature ceases to be a requirement during cooling, the cooler may be operated as a moving packed bed, as shown in Fig. 11.1. Adequate cooling of the hot wheat was effected in the moving bed at approximately half the air flow rate used in the drier, though a somewhat wider column had to be used—38 cm (15 in.) diameter as against 30 cm (12 in.) diameter for the drier—to allow sufficient residence time. Moisture removal in the cooler amounted to about $\frac{1}{3}$ of that in the drier. Up to 270 kg/hr of wheat, through a dry basis moisture range of 4%, could be dried in a system of this size, using 177°C inlet air.

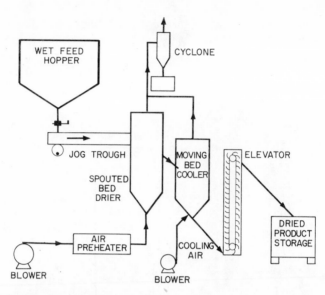

Fig. 11.2. Layout of an industrial grain drier for agricultural products [178]. Reproduced by permission of *The Canadian Journal of Chemical Engineering* **40**, 226 (1962), Fig. 2.

   The layout of a similar industrial installation consisting of a 61 cm diameter spouted bed drier (with a 1.78 m deep bed) and a 76 cm diameter moving bed cooler is shown in Fig. 11.2. Two such units have been in satisfactory operation in Canada for over ten years, for drying peas, lentils, and flax. The air for the spouter is heated by direct combustion of natural gas, and a steady bed temperature is ensured by the provision of a modulating valve on the burner, actuated by a temperature controller in the bed. This installation is capable of drying almost 2000 kg of peas per hour through an 8.8% moisture range (dry basis), though the capacity varies considerably for different types of peas (see Table 11.2). It occupies a floor area of 2.4 m × 7.6 m with a height of about 7.6 m exclusive of ductwork and cyclone. The capacity in terms of kilograms of water evaporated per hour per cubic meter of the drier (including the cooler, plus freeboard above both beds) varied between 24 and 112, which, according to Peterson [178], is very much higher than for conventional moving bed or cascade driers, and similar to that for a rotary drier (32–96 kg/hr/m³). On the economics of spouted bed drying, Peterson estimated that, while the capital cost of the plant in Fig. 11.2 was only about $\frac{1}{3}$ of an equivalent cascade drier, the operating cost of the two types was comparable (0.4 cent/kg of peas). The similarity in operating costs, despite the fact that the air consumption in a spouted bed drier is substantially lower, arises from the higher blower delivery pressure required in the spouted drier. Thus, in wheat drying, the power cost for a spouted bed drier, as worked out by Becker and Sallans [16], amounted to more than twice that for a conventional moving bed type of drier commonly used in North America. The overall picture which emerges from the above comparisons is that spouted bed drying offers large savings in capital cost and space over conventional drying methods, at no additional operating cost. It should be noted that for materials which tend to stick or form lumps during drying, and therefore cannot be easily handled in conventional driers, the advantage of using a spouted bed would become much greater. Examples of such materials in Table 11.2 are manganese chloride, ammonium nitrate, and gelatin granules, all of which tended to cake so badly that attempts to dry them even in fluidized beds were unsuccessful. For such materials, the breakdown of embryonic agglomerates in the high velocity spout constitutes a crucial advantage.

B. Drying of Suspensions and Solutions

1. *In a Bed of Seed Particles: Granulation*

The use of a spouted bed for producing millimeter size granules starting from pastes, suspensions, or solutions was first proposed by Berquin [19], and forms the subject of a patent issued to PEC of France [20] and of a later patent to ICI of Britain [167]. The bed in this process consists of particles of the material which is to be granulated (seed granules), and the liquid phase is injected into the base of the bed together with the hot spouting gas, as shown in Fig. 11.3. A thin layer of the liquid is deposited on the circulating particles as they pass through the liquid spray, which is dried by the action of hot gas as the particles travel further up the spout and down the annulus. Thus a particle builds up by a mechanism of layer-by-layer growth as it cycles in the bed, with opportunity for each layer to dry before the deposition of the next layer. The granule produced by this mechanism is well rounded and of uniform structure, since any irregularities are evened out as the particle builds up (see Fig. 11.4). The process lends itself to continuous operation and yields a product of uniform size with only small proportions of undersized or oversized materials.

Such a process can also be carried out in a fluidized bed [151; 211; 212; 242, p. 89], but the use of a spouted bed enables granules of a much larger size to be produced. Apart from this, the existence of a high voidage, high-gas-velocity zone at the bottom of the spout, which is also the hottest zone in the bed, provides an ideal location for injecting the liquid spray. Rapid evaporation of water in this region causes the temperature of the hot gas to fall sharply before it comes in contact with the bulk of the bed solids.

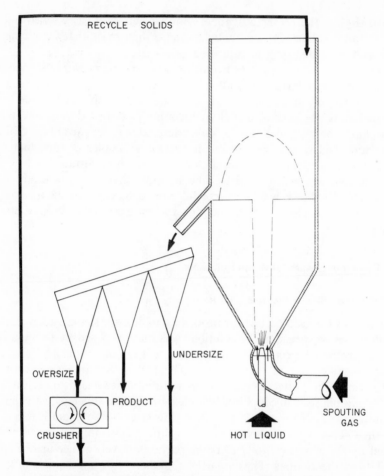

RECYCLE SOLIDS

UNDERSIZE

OVERSIZE

PRODUCT

CRUSHER

HOT LIQUID

SPOUTING GAS

Fig. 11.3. Spouted bed granulation system, after Berquin [20].

Higher inlet gas temperatures can therefore be used without causing thermal damage to the bed particles than would be permissible in a fluidized bed, with consequent improvement in evaporation rates. The most important advantage, however, arises from the systematic cyclic movement of particles in a spouted bed, as against the more random motion in a fluidized bed, since with the former, agglomeration is avoided, so that the granules produced are uniform in size and possess a homogeneous layered structure.

In comparing spouted bed granulation with other conventional granulation techniques, Berquin [20] has pointed out that granulation, whether in a rotary drum or in a system where the liquid phase is sprayed into a stream of hot gas, yields a fine product containing a wide range of particle

Fig. 11.4. Spouted bed granulated product: (a) Sulfur, (b) nickel, (c) urea, (d) ammonium sulfate. Samples (a), (b), and (c) provided by Berquin [20]; sample (d) recently produced at UBC by Uemaki [239].

size. For obtaining large granules of uniform size, the prilling process is usually employed, but this can be used only with solutions of very low moisture contents. If the moisture content is more than a few percent, the height of the prilling tower required becomes prohibitive. In any case a prilling tower, being commonly 30 m or more tall, represents a large capital expenditure and is therefore economically viable only for large capacities. Even with such heights, the prills produced sometimes require further drying using other equipment, such as rotary driers, to remove the last few percent of moisture from the product. Spouted bed granulation, on the other hand, places no restriction on the moisture content of the liquid phase, requires compact equipment, and is capable of producing a dry product in one step.

With highly concentrated solutions such as are normally processed in a prilling tower (e.g., in the manufacture of ammonium nitrate and urea), cold rather than hot air can be used in the spouted bed system, with consequent thermal economy. The melt is preheated to a temperature above its crystallization point and crystallization commences during atomization of the melt as a result of rapid cooling of the liquid droplets by the action of the high velocity cold air stream. The subsequent process of particle buildup and drying is the same as described above. Another variation of the process

Table 11.3   Spouted Bed Granulation Data for Some Agricultural Chemicals[a]

| Material | Feed solution | | Product | | | Air temperature (°C) | | Air flow rate (m³/sec) | Capacity (Mg/hr of product) | Weight of bed of seed granules (Mg) |
	Moisture (%)	Temperature (°C)	Size (mm)	Moisture (%)		Inlet	Outlet			
Complex fertilizer (nitro-phosphorus)	27	Cold	3–3.5 (90%)	2.4		170	70	13.9	4	–
Potassium chloride	68	Cold	4–5 (oversize <5%)	–		200	60–65	13.9	1	1
Ammonium nitrate	4	175	2.5–4	0.2		Cold	55	13.9	9.5	1.5
Sulfur	0	135	2–5	0		Cold	–	0.011[b]	0.04	0.008

[a] Performance data reported by Berquin [20].
[b] Injecting 1 liter/hr water as spray into the spouting air reduced the air requirement to 0.007 m³/sec for the same product output.

TABLE 11.4 Spouted Bed Granulation Data for Some Selected Materials[a]

Material	Feed liquor (water content %)	Product Size (mm)	Product Moisture (%)	Air temperature (°C) Inlet	Air temperature (°C) Outlet	Bed size D_c (mm)	Bed size H (cm)	Air usage (kg/kg moisture removed)	Coefficient of heat utilization (%)
Inorganic pigments [p. 205][b]									
Natural sienna	45.2	3–5	0	280	100	87	22	15.8	57.0
Blanc fix	25.5	2–4	0.03	352	95	30	25	19.2	60.6
Whitewash	90.0	1–3	8.2	460	85	20	32	9.8	76.5
Organic dyes [p. 206][b]									
Direct black 3	69.0	2–4	6.0	390	115	30	28	11.5	52.0
	68.2	2–4	6.5	310	93.5	160	40	10.7	70.0
Acid blue black	63.0	1–3	6.5	226	154	30	27	83.0	11.6
Other substances [p. 216][b]									
Sodium trichloroacetate	50.0	3–4	1.0	170	82	20	20	39.2	43.5
	50.0	3–4	3.3	175	57	90	50	35.0	50.0
Calcium chloride	65.7	3–6	5.6	380	126	90	55[c]	10.6	61.5
Sodium metadisulfate of benzol	50.0	<5	0	380	160	160	40	21.0	30.0

[a] From the extensive results of Romankov and Rashkovskaya [201].
[b] Page numbers refer to the Romankov–Rashkovskaya book [201].
[c] Estimated from bed weight.

covered by the Berquin patent concerns the granulation of materials which are normally in solid form at ambient temperature but which can be melted without deterioration of properties. Examples of materials quoted are metals such as lead and bismuth, their alloys, sulfur, and organic substances like naphthalene. The temperature of the spouting gas (which may be an inert gas such as nitrogen, if necessary) is kept below the melting point of the material to be granulated, which is premelted and introduced into a bed of seed granules together with the spouting gas, as before. The atomized droplets cool down to a temperature close to the solidification point in the lower part of the spout but actual solidification does not occur until the droplets are deposited on the bed particles in the upper, low voidage, part of the spout.

Some performance data reported by Berquin for each of the above three types of applications are given in Table 11.3, while a sketch of the granulating system used is shown in Fig. 11.3. Included in the sketch is the arrangement for recycling of undersized and crushed oversized material. Berquin has, however, noted that practically no oversized granules were produced, while the proportion of undersized particles in the overflow from the granulator could be kept to a minimum by "judicious location of the overflow pipe." This effect was made possible by a certain degree of classification which reportedly occurred, with the large granules segregating out in the upper part of the annulus near its periphery.

Further data on spouted bed granulation of fertilizers in 15 and 160 cm diameter units with bed depths of 36 and 152 cm, respectively, are reported in the ICI patent cited earlier [167]. Production rates quoted are in the 2–3 kg/hr range for the smaller unit and 200 kg/hr for the larger. Additional features covered by this patent include methods for avoiding blockage of the liquid spray nozzle, introducing a part or all of the recycle material by entraining it in the spouting gas (as in Fig. 1.3b), using a mixture of ammonia and air as the spouting medium for products containing diammonium phosphate, and methods for preparing complex fertilizer slurries suitable for use in the spouted bed granulator.

The technique of spouted bed granulation has proved to be useful for even a wider range of solid materials than those cited by Berquin. Extensive data on granulation of materials such as inorganic pigments, organic dyes, and a number of other heat-sensitive substances have been reported by Romankov and Rashkovskaya [201], for both bench scale and industrial sized units. A selection from their results is shown in Table 11.4, and the arrangement of one industrial installation designed by these workers in Fig. 11.5. The system used by the Soviet workers is essentially the same as that of Berquin, but they experimented with different feeding arrangements

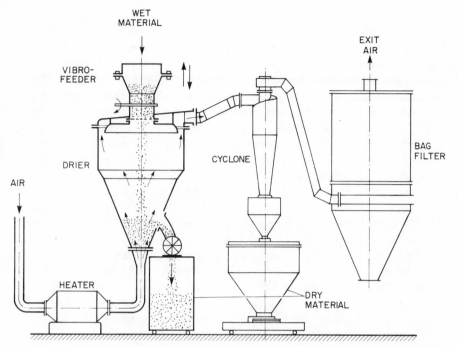

Fig. 11.5. An industrial spouted bed installation of Soviet design for granulation (or drying) of pasty materials. (Reproduced from Romankov and Rashkovskaya [201, p. 203].)

and found that for viscous pasty materials, introducing the feed into the top of the bed through a vibrating feeder (as in Fig. 11.5) worked well. For high-moisture-content solutions, on the other hand, injecting the feed into the high-temperature lower region of the bed, either coaxially with the spouting air as in Fig. 11.3 or sideways through the column wall, proved more suitable. It should be noted that although the spouting columns used by Romankov *et al.* in their granulation experiments were of cylindrical-conical shape, the operating bed depths were so small (see Table 11.4) that in most cases the bed would have remained within the lower conical part of the apparatus, unlike Berquin's granulator in Fig. 11.3, the overflow pipe location of which shows that the bed extended to a considerable height into the cylindrical part.

On the mechanism of the process, Romankov and co-workers have noted that side by side with the growth of the original bed granules, a certain number of new particles are constantly formed within the bed, partly by breakdown of the growing particles due to attrition and partly by evaporation of those liquid droplets which fail to deposit on existing particles.

These newly formed particles, together with the fine recycled particles, serve as nuclei for the formation of new granules, although a certain proportion of these fine particles is elutriated out of the bed before they have an opportunity to grow. Steady state operation is therefore possible only when the rate of formation of granules of the desired size range is so balanced by the net rate of appearance of fresh nuclei in the bed that the size distribution of the bulk bed material remains substantially unchanged with time and within the limits of spouting stability. Thus, the key operating variables are not only the temperature of the inlet air and feed rate of the liquid phase which would control the growth rate, but also the size distribution and rate of recycle of fine solids.

A generalized treatment for predicting the performance of such a system has been proposed by Kunii and Levenspiel [110] in the context of fluidized beds, using the basic assumption of backmix flow of solids in the bed. Since this assumption is also valid in the case of a spouted bed, the same theoretical analysis should apply here. The equations developed [110, Eqs. 23 and 26, Chapter 11] enable the steady state flow rates and size distributions of the solid streams, namely, product, recycle, and elutriated fines, to be calculated, provided that the rate of particle growth and of elutriation are independently known from laboratory scale experiments. Other equations for fluidized bed granulation, based on a similar approach, are given in the Romankov–Rashkovskaya book [201, Chapter V], some in conjunction with empirical expressions relating growth rate and elutriation with hydrodynamic and thermal parameters of particular systems.

An experimental study of the kinetics of particle growth in a spouted bed granulator with no solids recycle has recently been carried out at UBC using ammonium sulfate [239]. The net growth rate under a given set of operating conditions was found to remain independent of the (Sauter) mean diameter of the growing bed particles (up to a mean diameter of 3.4 mm, which was the maximum reached in the experiments), and to increase in approximate proportion to the solution feed rate raised to the power 0.8. The rate of particle attrition under granulating conditions, measured in separate experiments, was also found to be substantially independent of mean particle size, but it varied widely depending on the average moisture content and temperature of the bed material. The attrition data obtained so far are too few for developing any quantitative relationship between attrition rate and the abovementioned bed conditions. It does, however, appear that factors such as thermal stresses within a granule and moisture removal rate have a strong influence on the physical structure of the granules and their attrition behavior.

2. *In a Bed of Inert Solids*

An ingenious method for drying of solutions (evaporative crystallization) and suspensions in a bed of inert particles of spoutable size, e.g., glass beads of 3–6 mm diameter, has been developed at the Leningrad Institute of Technology [193, 154, 155, 201] for applications where the dried solids are ultimately required in the form of a fine powder rather than as large granules. The solution, containing up to 85% water, is atomized into the lower region of a bed spouted with hot air, deposits as a thin film on the glass beads passing through that region, and subsequently dries, as in the granulation process described in the previous section. The film, however, does not continue to grow on the glass beads during successive cycles; it becomes progressively fragile as it gets drier and is knocked off the particle surface by interparticle collisions in the spout. With the relatively large and heavy inert particles used, the film is reduced to a fine powder during the collisions, elutriated out of the bed by the spouting air, and collected as the product in an overhead cyclone.

The cyclic process of deposition, drying, and demolition of the surface film proceeds continuously provided that the bed temperature for a given feed rate of solution is maintained at a sufficiently high level so that the proportion of wet and sticky particles in the bed under steady state conditions remains small, and the spouting movement in the bed as a whole is not jeopardized by agglomeration of the wet particles. Stability of the process depends not only on the rate of drying of the wet film but also on the mechanical properties of the dried film, since if this film were to adhere tenaciously to the particle surface, it would simply continue to grow as in granulation. The minimum moisture content attainable also depends on the attrition characteristics of the dried surface film, since the attrition rate, in addition to being dependent on hydrodynamic factors, such as air velocity and size and density of the inert particles, is also a function of moisture content. For instance, if the surface film of a particular material becomes very fragile when its moisture content reaches $x\%$, then the process would obviously not permit drying that material to less than $x\%$ moisture. Since the mechanical properties of a surface film and its dependence on moisture content are specific properties of a material, the suitability of spouted bed drying for a given application and the optimum size and density of the inert particles must be determined by experiment.

The method described has been successfully used by the Leningrad group for drying a wide variety of materials, including organic dyes and dye intermediates, lacquers, salt and sugar solutions, and various chemical reagents.

Detailed data for both laboratory scale and industrial driers can be found in the Romankov–Rashkovskaya book [201, Chapter IV]. Included in the book is a comparison of the actual costs incurred in the Soviet Union in drying a black dye, using different types of driers, at an industrial plant of 6000 kg/24 hr capacity [201, Table IV-16]. The total drying cost per 1000 kg of dried dye containing 5% moisture, starting from a 65% moisture paste, is quoted as 24.3 rubles for the spouted bed drier as against 46–48 rubles for a double-drum drier of the same capacity.

Reger *et al.* [193] have emphasized that close control of the exit air temperature (or bed temperature) is essential to the success of the drying process; if it falls below a certain minimum value, the glass beads tend to stick together, disrupting the spouting action, while an upper limit on it is placed by consideration of thermal damage to the dried product. The results of experiments on three different azo dyes using various inlet air temperatures (100–180°C), bed heights, and air flow rates showed that the exit air temperature T_{ge} could be approximately related to the final moisture content (wet basis) of the dried product m_e' by the following equation:

$$T_{ge} = 109 \exp(-0.0435m_e') \qquad (11.1)$$

The observed values of T_{ge} were in the range 60–100°C and those of m_e' were 2–11%. While the validity of Eq. (11.1) is limited to the data on which it is based, a direct dependence between the product moisture content and the exit air temperature regardless of the other operating variables was found to exist generally, and was subsequently confirmed in a series of statistically designed experiments on four different dyes, two of organic and two of inorganic origin [154, 155]. An empirical equation correlating some of the drying data for organic dyes has been developed on the assumption that the overall process is controlled by the rate of drying of the surface film rather than by the rate of its attrition [201, p. 272]. It was also assumed that the drying rate in turn is controlled by the external resistance rather than by internal diffusion; that is, drying occurs in the constant rate period. With the selection of dimensionless groups relevant to convective transfer of heat and including bed height through the parameter H/d_p, it was proposed that

$$\text{Nu} = f(\text{Re, Gu, Ar, } H/d_p) \qquad (11.2)$$

The heat transfer coefficient h_p, on which the Nusselt number is based, is defined as $q/A_p\,\Delta T$, where q is the quantity of heat required per unit time for evaporation of moisture, A_p the surface area of inert particles in the bed, and ΔT the logarithmic mean between the gas and solids temperature differences at the inlet and at the exit of the drier, the solids temperature being taken as the wet bulb temperature of the inlet air. The Reynolds number is based on the air velocity through the orifice, while the Gukhman number, Gu, is defined as $(T_{ge} - T_{gw})/T_{ge}$, T_{gw} being the wet bulb temperature of the inlet air. The experimental data, which included variation in the size of the glass beads constituting the bed (2, 3, 3.5, and 5 mm), were correlated by the following equation:

$$\mathrm{Nu} = 0.0597 \mathrm{Ar}^{-0.438} \mathrm{Re}_i^{2.0} \mathrm{Gu}^{0.61} (H/d_p)^{-1.0} \tag{11.3}$$

the range of values of the dimensionless groups covered by the data being Ar = 0.31–5.0 × 10⁶, Re_i = 935–1700, Gu = 0.18–0.292, and H/d_p = 40–55. The above equation, based as it is on a fairly narrow range of data, cannot be expected to have any general validity, so that the design of a drier of this type must rely on pilot plant experiments with the particular material. The form of the equation might nevertheless be more generally useful for interpreting and scaling up of pilot plant data, provided that the rate-controlling step is drying rather than attrition of the dried film.

The use of a spouted bed of inert particles for predrying lime mud from the chemical recovery system of a kraft pulp mill, prior to the calcining step, has been proposed recently [161]. The primary objective of the proposed scheme is not simply drying but odor control, by suppressing the formation of H_2S arising from the sulfide present in lime mud. Experiments carried out using a 15 cm diameter × ~40 cm deep bed of −4 +10 mesh cement clinker particles have shown that with the intimate gas–solids contact in a spouted bed, the sulfide is almost fully oxidized, and that the drier can be operated continuously at high capacities, the comminutive action in the spout being effective in demolishing agglomerates. The economics of spouted bed drying followed by calcining in a conventional kiln appear to be particularly attractive for mills planning future expansion.

C. The Spouted Bed as a Reactor–Granulator

The possibility of using a spouted bed as a combined reactor–granulator, as well as a drier, has been explored by two independent groups of investigators, at Tashkent in the USSR [255, 229, 232] and at Trail in Canada [41]. Both groups initially tried using a fluidized bed for carrying out the combined process, as had been previously done by Lutz [123], but found that the use of a spouted bed yielded a rounder, smoother, and harder granule without the problems of agglomeration and scaling of the distributor plate, as well as of reactor walls, encountered in fluidized bed operation. Photographs of mixed-fertilizer granules produced by the Canadian workers

Fig. 11.6. Comparison of ammonium phosphate fertilizer produced in pug mill, fluidized bed, and spouted bed. (a) Usual plant product from pug mill; (b) product from laboratory fluidized bed; (c) product from laboratory spouted bed. (Courtesy B. McDonnell, Cominco [41].)

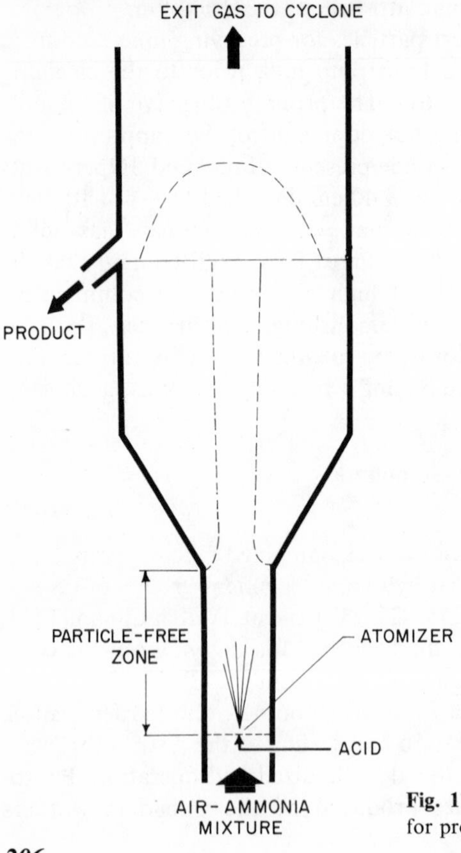

EXIT GAS TO CYCLONE

PRODUCT

PARTICLE-FREE ZONE

ATOMIZER

ACID

AIR-AMMONIA MIXTURE

Fig. 11.7. Reactor–granulator of Vyzgo *et al.* for producing nitric–phosphate fertilizer [255].

under both fluidizing and spouting conditions are shown in Fig. 11.6. A photograph of the same product manufactured in a conventional blunger (pug mill) granulator is included in the figure for comparison.

Vyzgo and co-workers [255] experimented with producing nitric–phosphate fertilizer mainly in a 32.5 cm diameter reactor. The bed consisted of product granules (1.5 mm size) and was spouted with a gaseous mixture of air and ammonia at 100–120°C. A preheated acid solution, obtained by the nitric acid decomposition of rock phosphate and containing 25–50% water, was atomized into the spouting gas a few centimeters below the point of its entry into the bed (see Fig. 11.7). A grate with about 50% open area was placed in the gas inlet pipe at the level of the liquid injection, so that the linear velocity immediately above the grate, at the gas flow rates used, became sufficiently high to ensure that the section of the pipe between the grate and the narrow end of the cone remained substantially free of particles. This arrangement, according to the authors, is a key feature of the process, since it enabled neutralization, which is accompanied by rapid evaporation of moisture, to be completed before the products of reaction made contact with the bed solids. The subsequent granulation process within the bed occurs in the same way as described in the previous section. The bed temperatures observed in these experiments were 60–70°C, and the product moisture contents less than 1% [255]. The authors have claimed that the fertilizer produced contained a higher proportion of assimilable phosphates (90% of the total phosphates) than is commonly achieved. Formation of undesirable phosphates during ammoniation was avoided by the intensive mixing in the neutralization zone, while deterioration caused by local overheating during the granulation step was minimized by virtue of the well-mixed nature of the solids.

The Canadian work cited above was primarily concerned with ammoniation of phosphoric acid to produce ammonium phosphate fertilizers, and was carried out in a 14 cm diameter column using bed depths of 2–3 column diameters. Acid concentrations tested were 23% and 40% P_2O_5, with ammonia to air ratio in the spouting gas in the 5–10% (by volume) range. Bed temperatures recorded under these conditions varied between 50 and 65°C. While the importance of proper acid distribution has been emphasized by these workers too, they did not adopt any special measures to keep the neutralization zone free of solid particles, as did Vyzgo *et al*. Instead, the acid was atomized vertically upward through a central nozzle located concentrically with the gas inlet pipe and projecting slightly into the bed. Such an arrangement appears to have proved quite satisfactory, at least for the limited range of experimental conditions explored.

D. Particle Coating

1. Deposition from a Liquid Solution

The principle of spouted bed granulation can also be used for coating of particles and has been successfully applied at Abbott Laboratories by Singiser and co-workers [91, 216, 217] in the pharmaceutical industry to tablet coating, where the requirements for uniformity of coating thickness are particularly stringent. The arrangement for feeding the coating liquid into the spouted bed was similar to that used by Berquin (see Fig. 11.3), but the operation was carried out batchwise, presumably because this ensured equal residence time in the bed for the individual particles, production rates involved in tablet manufacture being relatively small in any case. The actual coating unit which coats batches of 70–100 kg of tablets is shown in Fig. 11.8 [217]. Its operation consists in loading a batch of the tablets to be coated into the column, turning on the hot air supply to spout the bed, and then starting the flow of the preheated coating liquid through the pneumatic atomizing nozzle. The rate of liquid flow is so regulated that

Fig. 11.8. An industrial tablet coating unit in operation [217]. Picture on the right shows finished tablets in one column; the second column is loaded and ready to be positioned over plenum for coating.

the spouting action is not impaired due to stickiness caused by excessive surface moisture on the tablets. After the desired quantity of coating solution has been supplied to the bed, a period of drying to remove any residual solvent from the coating is allowed at a reduced air flow rate with the bed in a quiescent condition. Since drying of the solution during the coating operation occurs almost instantaneously, there is little danger of solvent penetration into the tablet core, and therefore the final drying of the coated batch takes only a few minutes. Coating times required for 70–100 kg batches were between 1 and 1.5 hr, the inlet and exit air temperatures being typically 63°C and 26°C, respectively [91].

The results of film thickness measurements made by Singiser and Lowenthal [216], which are reproduced in Table 11.5, demonstrate the uniformity

TABLE 11.5 Film Thickness Data for Spouted Bed Coating

Tablet size (mm)	Shape	Surface area (mm²)	Tablet weight (mg)	Tablet face		Tablet wall	
				Average thickness[a] (μm)	Range (μm)	Average thickness[a] (μm)	Range (μm)
10	Standard concave	285	500	107	89–133	97	97–114
10	Deep cup	259	500	136	114–159	109	82–133

[a] Average of 25 micrometer readings.

of coverage obtained in spouted bed coating. Good batch to batch uniformity was also achieved, since once proper operating conditions (namely, charge weight, flow rate and temperature of air, and feed rate and temperature of the coating liquid) had been established for each product, close control of these could be easily exercised. The workers at Abbott Laboratories have reported that tablets coated by this process consistently met the pharmaceutical standard (U.S.P. XV enteric test), thus eliminating the need for recoating and subsequent retesting. A similar industrial unit is known to be in satisfactory operation in Canada [65].

Singiser *et al.* [217] have noted that the problem of tablet attrition, which is an obvious disadvantage of spouted bed coating, has been substantially solved to the point that no inspection or sorting of the coated tablets is necessary. This appears to have been achieved by reformulation of soft or friable tablet cores, or by first seal-coating such tablets using conventional coating methods before subjecting them to spouted bed coating. The absence of a polishing action in the bed is mentioned as a second disadvantage of the technique, subsequent polishing of the tablets in a conventional pan being necessary in certain cases. Nevertheless, the short coating time in the spouted bed process gives it an overriding economic advantage since, in general, a spouted bed unit can replace as many as 10–12 conventional coating pans.

Examples of other commercial operations for which the coating technique described above should be well suited are encapsulation of pelletized artificial fish food to make it water-resistant, and of seed grain to delay germination. Some preliminary experimental work on the former application has been carried out at the Univ. of Brit. Columbia [93]. Coating of granular fertilizers and other chemical substances to protect them against caking in storage or to reduce their rate of solution are further possibilities, mentioned by Nichols [167].

2. Thermochemical Deposition

Several investigations on coating of uranium oxide and uranium carbide particles with pyrolitic carbon, in an apparatus of the type shown in Fig. 11.9, have been reported in recent years [14, 25, 24, 1, 3], mainly from the USA. Only small coating units (less than 50 mm dia) have been used with shallow beds ($H/D_c < 2.5$) of uniform size particles in the size range 0.2–0.6 mm. While the experimenters themselves have described the process as "fluidized bed coating," Abdelrazek [1], who made a detailed study of the hydrodynamics of beds similar to those used in the coating experiments, has identified the coating unit as a spouted bed system.

The use of carbon-coated nuclear fuel particles holds potential for high-temperature, gas-cooled convertor reactors, but the technical feasibility of this concept depends on the ability of the coating to retain its structural integrity during reactor service [14]. Since failure can arise from a large number of causes, such as fuel swelling, fission-gas pressure, thermal expansion or contraction of coating, radiation damage, or chemical reaction of fission products with inner coating surface [184], the properties of the coating required, which depend on the microstructure of the deposited carbon, are highly specific. The investigations using spouted beds have therefore been mainly concerned with determining the structural characteristics of carbon coatings formed under varying deposition conditions, e.g., of bed temperature, the particular hydrocarbon used (methane, propane, acetylene, etc.), its concentration in the spouting gas, gas flow rate, and the surface area of the bed solids.

An extensive review of this work has been presented by Abdelrazek, who has also proposed a simplified model to describe the process of pyrolytic carbon deposition in a spouted bed. The model is based on the assumptions that the bed is isothermal, that the coating takes place mostly in the annulus, and that the coating rate is controlled by the rate of diffusion of some

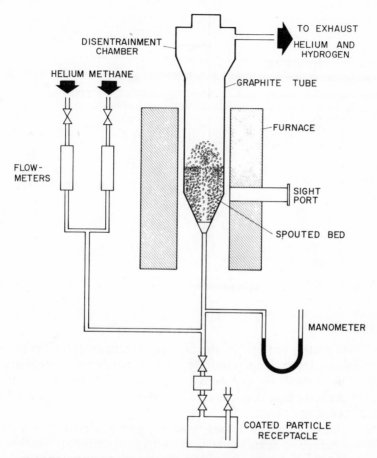

Fig. 11.9. Schematic diagram of a spouted bed pyrolytic carbon coater [1].

hydrocarbon species (either the parent hydrocarbon or an intermediate compound) from the gas phase to the solid surface. Mass transfer coefficients calculated on this basis from the experimental results of Beutler and Beatty [24] on overall deposition rates from methane onto uranium carbide particles of 0.53 mm diameter (coating thickness, 70 μm), when plotted against the reciprocal of the absolute bed temperature, yielded an activation energy value of 12.95 kcal/mole for bed temperatures between 1500 and 1800°C. The activation energy for temperatures below 1500°C (down to 1300°C) was found to be much higher. The author therefore concludes that mass transfer does control the deposition process at least down to 1500°C while at lower temperatures the process is controlled by the kinetics of the pyrolysis reaction.

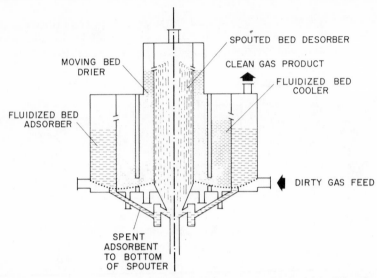

SPOUTED BED DESORBER

MOVING BED
DRIER

CLEAN GAS PRODUCT

FLUIDIZED BED
COOLER

FLUIDIZED BED
ADSORBER

DIRTY GAS FEED

SPENT
ADSORBENT
TO BOTTOM
OF SPOUTER

Fig. 11.10. Continuous adsorption equipment ("Ecosorber") developed by Németh *et al.* [165].

The same coating process has also been considered at Harwell in Britain, but the only unclassified report on it [36] is concerned with the hydrodynamic behavior of shallow spouted beds consisting of 0.5–6.4 mm diameter spherical particles with densities up to 11 Mg/m³, contained in a 7.6 cm diameter vessel (see Chapter 2).

The choice of a cone-bottomed vessel for pyrolytic coating is dictated not by any particular desire to obtain spouting in preference to fluidization, but rather by the necessity of avoiding the use of a distributor plate, since deposition on the plate itself can cause its blockage [36, 3]. Nevertheless, the resulting spouted bed appears to serve the process well.

E. Sorption

A continuous adsorption setup ("Ecosorber") for gas cleaning developed by Németh *et al.* [165] makes use of a spouted bed for carrying out the desorption step in the process, presumably with steam as the spouting fluid, while a fluidized bed is used for adsorption. The spent adsorbent (e.g., activated charcoal granules) is fed into the base of the desorber by entrainment into the spouting gas, and the regenerated particles from the top of the spouted bed overflow into the surrounding moving bed drier (see Fig. 11.10). Apart from the intensive gas–solid contact provided by the spouting action in the desorber, the choice of a spouted bed in this case is dictated by the particular configuration of the equipment, which requires elevation of the solids from the bottom of the adsorber to the top of the drying compartment. Since the vertical transport of solids was achieved at one-tenth the gas velocity for pneumatic conveying, a considerable saving in power cost was realized by combining the desorption process with vertical transport through the use of a spouted bed, instead of transporting the solids pneumatically as for a hypersorber. The novelty of the system as a whole lies in its compactness, achieved by arranging the adsorption, desorption, drying, and heating zones in the form of concentric cylindrical compartments.

F. Potential Diffusional Operations

There are other diffusional operations, or variants of the above, which suggest themselves as possible candidates for spouted beds, but which have yet to be tested. Vapor impregnation of porous solids, e.g., steaming of wood chips, is the converse of solids drying and could presumably be carried out by using the vapor as the spouting fluid. This operation, as well as the simpler adsorption used in gas cleaning, is suggestive in its execution of gas dehumidification by hydration of coarse anhydrous solids such as calcium chloride or calcium sulfate granules, and also of desublimation by buildup to coarse solids such as naphthalene or p-dichlorobenzene "moth balls." On the other hand, removal of a low volatility liquid (e.g., an oil) from the pores of granules could possibly be accomplished by partial pressure distillation (e.g., steam distillation) using hot entraining vapors (e.g., superheated steam) as the spouting fluid, supplemented if necessary by internal or jacket heating.

In each of these operations, as well as any others that might be devised, the inherent features of spouted beds, such as high jet velocities, good solids recirculation, and continuous operation, would only justify the commercial use of a spouted bed if these advantages were not eclipsed by some of the disadvantages, such as attrition of solids, moderately high power consumption, nonuniformity of particle residence times, etc. It is an engineering commonplace that technical feasibility of a given process does not necessarily imply its practicability on a large scale without comparative assessment against alternate means of performing the same function.

11.3 THERMAL

Heating or cooling of coarse granular solids without concomitant mass transfer may be carried out in a spouted bed, in much the same way as drying or adsorption, by transferring heat from or to the spouting gas. The full benefit of the technique is, however, realized only in the case of heating of thermally sensitive particles, since the well-stirred nature of the bed permits the use of higher hot gas temperatures than would otherwise be possible without thermal damage to the particles. For continuous heating of solids, higher heat transfer rates can therefore be achieved than in a nonagitated system. In a cooling operation, on the other hand, the mixing of solids in the bed is not of critical value, but the ease with which solids can be fed into and discharged from the bed, together with the intimacy of

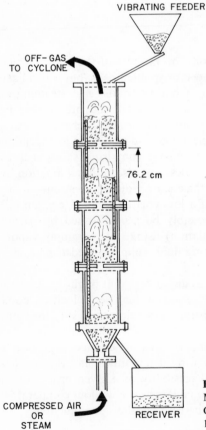

Fig. 11.11. Multistage spouted bed unit of Malek and Walsh for preheating coal [130]. Column diameter 15.2 cm; orifice diameter 1.27 cm.

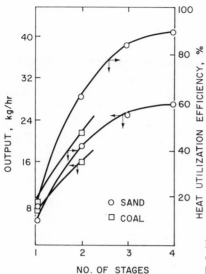

Fig. 11.12. Performance data for stagewise operation obtained using apparatus of Fig. 11.11 [130].

gas–particle contact, still make the spouted bed process a strong candidate for cooling of coarse solids.

A. Solids Heating

The possibility of preheating coking coal (6 mm particle size) in a bed spouted with hot air or steam as a means for increasing the output of coke ovens has been experimentally investigated by Malek and Walsh [130] with promising results. In continuous runs carried out on a 15 cm diameter unit, coal temperatures up to 250°C were achieved without agglomeration, using inlet fluid temperatures up to 350°C. These workers also experimented with a multistage unit of the type shown in Fig. 11.11 not only with coal but also with sand, obtaining higher throughput rates and heat utilization efficiencies with multistage operation (see Fig. 11.12). On the basis of these results, Malek and Walsh consider that a two- or three-stage operation would be the most efficient, the effect of increasing the number of stages from three to four being small.

Some further points noted by these investigators in connection with this particular application are:

(1) Multistage spouted bed operation should minimize the hazard due to explosion initiated by hot coal dust, since fine dust from the incoming coal is removed by elutriation in the topmost unit where the operating temperature is the lowest;

TABLE 11.6 Typical Operating Data for Fisons Double-Deck Multiple-Spout Fertilizer Coolers[a]

Deck size (m × m)	Capacity (Mg/hr)	Solids temperature (°C)		Air flow			Inlet air temperature (T_{gi}) (°C)	Thermal efficiency[b]
		Feed (T_{si})	Product (T_{se})	Volumetric (m³/sec)	Superficial velocity (m/sec)			
4.9 × 1.8	28.2	120	40	19.8	2.2		25	84
4.9 × 1.8	22.7	100	30	19.8	2.2		25	89
4.3 × 1.2	12.7	70	30	8.0	1.5		25	93

[a] Courtesy of Dr. J. A. Storrow, Levington Research Station, Fisons Ltd., Ipswich, Suffolk, England [62].
[b] Thermal efficiency = $(T_{si} - T_{se})/(T_{si} - T_{gi})$.

(2) Addition of oil or other bulk-density-controlling liquid could be easily and effectively done during preheating by spraying the liquid into the spout zone (as in the particle coating process described earlier in this chapter);

(3) It was found possible to heat the bed solids electrothermally by installing an internal screen at the interface between the spout and the annulus, to serve as an electrode charged relative to the shell of the spouting vessel.

B. Solids Cooling

The design of a large capacity cooler based on the spouting principle has been developed by Fisons Ltd. for cooling granular fertilizer [62]. A multispout system is used in a large rectangular vessel with a perforated plate as its base. The flow rate of cooling air, which enters the bed through the perforated base, is sufficient to induce localized spouting action in the solids above each perforation, giving rise to a multitude of spouting cells, each with its own upward moving dilute phase spout surrounded by a downward moving aerated dense phase annulus. The bed in this condition differs from the multiple-spouted bed of Peterson (described in Chapter 12) inasmuch as no physical partition exists between the individual spouting cells, and would appear to be rather similar to a normal fluidized bed. However, at the operating air flow rate, the multiple-spouted bed described above is considerably more dense and in more orderly movement than is usual in a fully fluidized bed [62] where, with the coarse particles involved, the air jet issuing from each perforation would transform into large bubbles within a short distance from the inlet point. The Fisons multispout cooler is arranged in a double deck, on the same lines as the two-stage system of Malek and Walsh for preheating of coal just described. Hot solids are fed into the upper stage and discharged from the lower, flowing from one to the other through an internal downcomer, countercurrent to the flow of the cooling air. Typical operating conditions for three such units which have been in industrial operation at the Fisons Fertilizer Plant at Immingham, England, are given in Table 11.6. Estimated thermal efficiencies are seen to be in excess of 80%.

Design equations for heating or cooling of solids have been presented in Chapter 8.

C. Food Processing

Cooking of foodstuffs, although it may involve both volatilization and chemical reaction, is probably often thermally controlled. The roasting of coffee beans has been accomplished in a spouted bed using superheated steam as the spouting fluid [180]. Since only a small amount of volatile material was driven over with the steam, it seems likely that hot air would be almost as effective as steam.

Other thermally controlled treatments of food should also lend themselves to successful execution in a spouted bed. One possible application is the freezing of peas, beans, corn, and other food particulates, either whole or diced, using refrigerated subzero air as the spouting fluid. Unlike fluidization in deep beds, where food processing usually involves immersion of the edible matter in a fluidized bed of inert solids [77], the corresponding spouted bed operation would be more adaptable to spouting of the relatively large grains of food themselves, providing they are sufficiently rugged to withstand the jet action of the gas. Although tray [64] or belt [185] freezers with shallow fluidized beds of large food particles are in commercial use, the difficulty with sticking of particles onto the gas distributor plate, which is sometimes encountered, would be eliminated by the use of a spouted bed. A further advantage should arise in the compactness of the equipment, since the residence time requirement in spouted bed operation could be met by employing a deeper bed but one which occupied much less floor area, compared to fluidized bed freezers.

11.4 MECHANICAL

In all the applications discussed so far, effective contact between the bed particles and the spouting fluid is of primary importance. The operations described in this section, on the other hand, rely entirely on interparticle contacts, brought about by mechanical agitation of the bed solids. For these operations, the fact that agitation is caused by the action of a gas jet rather than by some other stirring device is really incidental, the gas–particle interaction, which does of course occur, being only of secondary importance.

A. Solids Blending

The spouting technique has proved successful on the industrial scale for blending of polyester polymer [27, 95], which is initially produced in the form of small lumps or chips and requires subsequent blending to obtain improved uniformity in the spun fiber. The blending operation is carried out batchwise, the use of a spouted bed in preference to a mechanical blender being particularly advantageous for large batch sizes (above 57 m³ or 2000 ft³), when the mechanical design of the blender and of the drive transmission system become complex due to high power inputs. In addition, the need for providing a storage container separate from the more expensive mechanical blender is eliminated with spouted bed operation, since here the blender itself consists of nothing more than a simple storage vessel, connected for air supply.

The power consumption for spouted bed blending depends on the scale of operation, but in general it is less than for mechanical blending. According to Bowers *et al.* [27], power consumption per ton of solids to maintain a bed in the spouted state remains substantially constant as the diameter of the blender increases, whereas the time required for blending increases with increasing blender size. The total energy input per ton of material blended therefore increases with increasing scale of operation. Typical power consumption for spouting a bed of polymer chips has been quoted by these workers as 3.7 W/kg (5 hp/ton) with blending time varying between 10 and 120 min, depending on blender size. The corresponding total work input would therefore range between 2.2 and 27 kJ (kW-sec) per kilogram

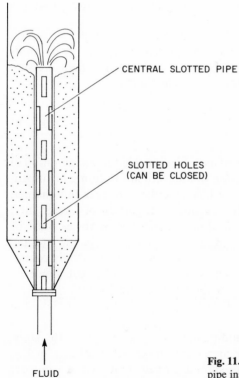

CENTRAL SLOTTED PIPE

SLOTTED HOLES
(CAN BE CLOSED)

FLUID

Fig. 11.13. ICI spouted bed blender with slotted pipe insert [27].

of solids blended. The comparable figures given for a mechanical blender are of the order of 10 W/kg and up to 100 kJ per kilogram of solids blended.

The existence of a peak pressure drop across the bed prior to the onset of spouting (discussed in Chapter 2) causes the power required for starting up a spouted bed to be considerably higher than that for operating it under steady conditions. In order to economize on the power consumption for starting up a spouted bed blender, Bowers *et al.* have patented the use of a vertical pipe in the axial region of the vessel, extending from the air inlet orifice to the bed surface or above. The pipe, of a diameter roughly $\frac{1}{8}$ of the column diameter, is provided with slots or holes spaced along its height, which can be closed by rotating a similar concentric slotted pipe so that the slots in the two pipes fail to coincide (see Fig. 11.13). The slots are closed initially while the bed is charged into the vessel and are opened only after the flow of air through the central pipe has been established. After the slots are opened and particles from the annular space can flow into the pipe, the solids flow pattern becomes much the same as in a normal spouted bed, that is, one without a central pipe. The restriction in the area of the spout–annulus interface is, however, bound to cause some reduction in the extent of solids cross flow. Other work on spouting with tube inserts is discussed in Chapter 12.

Extensive experimental data on the comparative performance of several types of solids mixers commonly used in industry, including an intermittently operated "spouting-bed type" of mixer with multiple high pressure air jets, have been reported by Ashton and Valentin [4].

B. Comminution

The concept of using a spouted bed as a grinding mill for comminution of particulate solids is a logical extension of the method for drying of suspensions and solutions in a bed of inert solids, described earlier in this chapter. The technique requires that the bed include a proportion of foreign inert particles to serve as the grinding medium. These inert particles must be harder and heavier than the active particles which are to be ground and must be capable of being spouted, that is, they should be relatively coarse, of uniform size, and free flowing. The same restrictions do not, however, apply to the active particles, since if their proportion in the bed is kept sufficiently small, the spouting performance of the bed as a whole is not adversely affected.

Experiments on spouted bed grinding have been carried out at the Univ. of Brit. Columbia [143] using the apparatus shown schematically in Fig. 11.14. The experiments involved air spouting a bed composed of a mixture

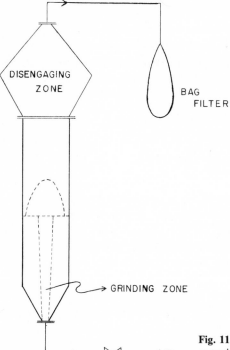

Fig. 11.14. Schematic diagram of experimental comminution apparatus [143].

of active and inert solids, and measuring the weight of the ground product collected in the bag filter as a function of time. Screen analyses carried out on the ground product and on the residual active material left behind in the bed showed only a small overlap of size range, indicating that the expanded section above the column provided efficient elutriation of the fine product as well as recycle of oversize material during the grinding process. The rate of elutriation was therefore taken as the effective rate of grinding. The majority of tests carried out being batch runs, the grinding rate tended to drop off with spouting time. The reported grinding rates were therefore based on the weight of fines collected over the first 10–30 min of operation, since it was assumed that these would be more representative of rates in continuous operation than those based on the cumulative weight of fines produced over a longer duration. Data from a few continuous runs, in which a stream of active particles was fed into the top of the bed to replenish the fines carry-over, confirmed this assumption.

Grinding rates for several combinations of active and inert materials were measured in preliminary experiments. The materials used and their properties are listed in Table 11.7, while the grinding performance of each

TABLE 11.7 Properties of Solid Materials Used in Spouted Bed Grinding Experiments

	Size range Tyler mesh	Particle specific gravity	Hardness (Mohs)
Materials ground (active solids)			
Urea	−4 +10	1.15	1–3
Cement clinker	−6 +10	3.08	5–6
Limestone	−6 +10	2.14	2–3
Gum arabic	−6 +10	1.51	2–3
Charcoal pellets	−6 +10	0.42	2–3
Grinding media (inert solids)			
Glass beads	5 mm diameter	2.39	6–7
Gravel	−4 +5	2.65	5–6
Ceramic balls	10 mm diameter	2.30	5–6
Lead shot	−7 +9	11.34	2–3

TABLE 11.8. Grinding Performance of Various Systems Based on Experiments Carried Out on a 15.2 cm diameter Spouted Bed Unit

Solid materials		Grinding performance[a]	Comments
Active	Inert		
Urea	Glass beads	Good	
	Gravel	Good	
	Ceramic balls	Good	
Cement clinker	Glass beads	Good	
	Gravel	Poor ⎫	Active and inert solids are
	Ceramic balls	Poor ⎬	of similar hardness.
	Lead shot	Poor	Active solids are softer and heavier than inerts; some segregation.
Limestone	Glass beads	Good	
Gum arabic	Glass beads	Fair	Only one test; grinding rate 2.4 kg/hr
Charcoal pellets	Glass beads	Poor	Large density difference caused segregation.

[a] Good: grinding rates greater than 4.5 kg/hr under favorable operating conditions; poor: grinding rates less than 1.5 kg/hr under favorable operating conditions.

system tested is qualitatively described in Table 11.8. In most cases the tests covered a range of such operating conditions as proportion of inerts in the bed, bed depth, and flow rate of spouting air.

The results in Table 11.8 show that

(1) A large density difference between the main solids and the grinding medium (charcoal with glass beads) led to segregation, and therefore to

poor grinding performance. No segregation was observed up to a density ratio of about 3 using similar sized solids.

(2) Inert solids of a hardness similar to or less than that of the main solids did not prove effective as the grinding medium, even when their particle size (ceramic balls versus cement clinker) or density (lead shot versus cement clinker) was considerably larger. This result points to the primary comminution mechanism as being one of abrasion rather than impact.

(3) With suitable choice of inert solids to serve as the grinding medium, and of operating conditions, the grinding rates are high enough to be of practical interest.

Further batch experiments to determine the effect of the main variables involved on grinding rate were carried out using only those pairs of solids which showed good grinding performance in the preliminary experiments. The data obtained are shown in Table 11.9, while the sieve analysis results for all the cement clinker runs are presented graphically in Fig. 11.15. Variations in the size distribution of the ground product obtained under different experimental conditions were not wide. This indicates that the disengaging section used (30 cm maximum diameter) above the spouting column (15 cm diameter) served as a consistent classifier.

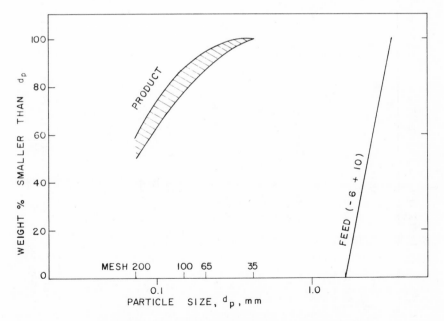

Fig. 11.15. Sieve analysis data for cement clinker [143].

TABLE 11.9 Grinding Rate Data from Batch Experiments Carried Out on a 15.2 cm Diameter Spouted Bed Unit[a]

Run no.	Materials		Wt% inert solids in bed (initial)	Bed size (initial)		Air flow rate (std. m³/hr)	Air inlet diameter (cm)	Grinding rate (kg/hr)
	Active	Inert		Height (cm)	Total wt. (kg)			
A. Effect of air inlet diameter								
7	Urea	Glass beads	50	27.9	3.4	87.7	1.27	6.3
41	Urea	Glass beads	50	33.0	3.6	100.1	1.90	2.2
35	Cement clinker	Glass beads	50	42.9	9.1	113.0	1.27	4.1
42	Cement clinker	Glass beads	50	43.9	9.1	112.3	1.90	2.2
B. Effect of air flow rate								
35	Cement clinker	Glass beads	50	42.9	9.1	113.0	1.27	4.1
36	Cement clinker	Glass beads	50	43.9	9.1	137.6	1.27	5.8
C. Effect of proportion of grinding medium in bed								
22	Cement clinker	Glass beads	0	50.0	10.3	77.1	1.27	1.5
21	Cement clinker	Glass beads	15	50.0	11.0	89.4	1.27	3.4
32	Cement clinker	Glass beads	30	50.5	10.6	106.0	1.27	4.4
31	Cement clinker	Glass beads	50	46.0	9.4	102.3	1.27	2.9[b]
D. Effect of bed depth								
28	Limestone	Glass beads	50	30.0	4.5	73.7	1.27	4.4
29	Limestone	Glass beads	30	50.0	8.2	75.1	1.27	10.4
20	Cement clinker	Glass beads	34	29.0	5.0	93.6	1.27	1.5
21	Cement clinker	Glass beads	15	50.0	11.0	89.4	1.27	3.4

[a] Operating air flow rates were roughly 15% above the minimum flow required for spouting (except in run number 36).
[b] Slugging occurred.

The following trends are suggested by the data in Table 11.9:

(1) The diameter of the air inlet orifice, which largely determines the jet velocity, is a critical parameter, a small decrease in orifice diameter causing a large increase in grinding rate. It would appear that there exists a threshold jet velocity which must be exceeded to cause appreciable breakdown of a given solid material.

(2) Grinding rate increases with increasing air flow rate, other conditions remaining unaltered. Raising the flow rate causes the spout to shoot up higher in the column. A baffle plate located in the axial region of the column was therefore used in run number 36 to restrain the spout height.

(3) For a given bed depth, grinding is favored by a high proportion of grinding medium, as long as the bed continues to spout smoothly. Too high a proportion of the heavy inert particles can, however, cause the spouting action to become sluggish, thereby affecting the grinding performance adversely (run number 31).

(4) Bed depth has a strong influence on grinding rate, the rates obtained in deeper beds being substantially higher even when the weight of inert solids in the bed is unaltered and therefore the proportion of inerts in the deeper beds lower. Here again, an upper limit on bed depth is set by the maximum spoutable bed depth. On approaching this limiting depth, the grinding rate was found to drop sharply.

Comminution in a spouted bed apparently occurs as a result of relatively low energy particle–particle interactions in the spout region. The grinding performance is therefore comparable with that of a conventional tumbling type of mill, rather than of high speed mills like hammer mills or fluid energy mills. Comparative tests with urea and cement clinker carried out in a laboratory ball mill and a rod mill showed that for a similar size reduction, spouted bed grinding rates were similar to those obtained in the tumbling mills. The energy consumption for comminution alone is estimated to be higher for spouted bed grinding, but in view of the fact that size classification is also achieved at no additional expenditure of energy, unlike the situation for tumbling mills, the energy consumption for spouted bed operation could become competitive. The more important advantage of spouted bed grinding over conventional methods, however, is considered to be the compactness and simplicity of the equipment, arising out of the absence of any directly driven mechanical parts. This should reflect itself in lower capital costs. The maintenance cost should also be substantially lower, particularly when grinding abrasive materials, since grinding action occurs in the hollowed central core of the bed and not against the vessel walls, thereby avoiding wear and tear of the main vessel. The maximum benefit of a spouted bed mill would be realized for operations requiring comminution simultaneously with effective gas–solid contact. Examples of such operations are grinding of hygroscopic or moist and sticky materials, of solids which tend to become plastic with rise in temperature (e.g., insecticides, resins, gums), of easily oxidizable or potentially explosive materials which must be ground in a completely inert atmosphere, and processes requiring chemical reaction between the solids and the gas simultaneously with grinding.

C. Potential Mechanical Operations

The particle attrition feature of a spouted bed, though undesirable for some operations, has turned out to be of critical value for others, namely, drying of suspensions and solutions on inert particles, coal carbonization, shale pyrolysis, iron ore reduction, and comminution (see Table 11.1). It therefore seems reasonable to assume that the dehusking of grain, and more generally the skinning of particles possessing a removable outer layer, could also be achieved by promoting controlled attrition during spouting.

Dust and mist collection from gases is another mechanical operation in which some of the special features of spouted beds could be exploited. The high velocity spout region should be effective in aerosol removal by inertial impaction and the annular region for removal by diffusional deposition. High gas throughput rates should be possible, since spouted bed operation would permit the use of coarse target particles, and continuous regeneration of these particles would be easy to arrange. The technique could have application to air pollution control as well as to high-temperature gas cleaning in conjunction with heat recovery.

Preliminary experiments to pursue this idea have been carried out recently at UBC [150]. With liquid aerosols of 0.1–2.7 μm size, collection efficiencies of up to 92% have been achieved, using a 15 cm diameter × 45 cm deep bed of cement clinker particles (-8 $+12$ mesh). Efficiencies obtained with solid aerosols, however, were substantially lower because aerosol particles failed to adhere permanently to the target particles and were consequently reentrained. Work to overcome the reentrainment problem is currently in progress.

CHEMICAL PROCESSES

Although a spouted reactor–granulator involves a chemical reaction, it has nevertheless been categorized in this chapter under Diffusional Operations (Section 11.2) because of its close affinity to evaporative physical granulation. Thermochemical deposition has been similarly categorized because of its close affinity to evaporative particle coating. On the other hand, low-temperature coal carbonization, though in some measure a thermal volatilization operation, has been classified here as a chemical process because of the important pyrolysis involved [215, p. 78]. The classification system adopted is thus somewhat arbitrary.

11.5 SOLIDS AS REACTANTS

A. Low-Temperature Coal Carbonization

Low-temperature carbonization of coal is a thermochemical process involving the evolution of volatile matter from coal in the approximate temperature range 450–700°C to form coke. Interest in low-temperature carbonization centers on the fact that the amount of tar produced, which is an important source of liquid fuel, is three to five times greater than in high-temperature (900–1175°C) carbonization [205, p. 400]. Low-temperature char is softer and more friable than the coke oven product produced at high temperatures, and is more suitable for use as a smokeless fuel and as a blend component for certain metallurgical purposes. In recent years, the main incentive for the development of the low-temperature process has come from its adaptability to continuous operation through the use of a fluidized bed where the heat necessary for endothermic carbonization is more rapidly, efficiently, and uniformly transferred to coal than in a coke oven, especially if the heat is supplied by partial combustion of the char within the bed itself [190, 29]. The use of a fluid–solid system at high temperature is ruled out by a severe caking problem, caused by an increase in the plasticity of coal particles when subjected to a high rate of temperature rise. The problem arises even during low-temperature carbonization if the particles are fine. Thus in a fluidized bed carbonizer, with coal particles in the 30–200 mesh range, it becomes necessary to operate with a large char recycle to avoid agglomeration—recycle ratios (char/coal) as high as 10/1 have been recommended for certain types of coal [112].

It was to avoid the agglomeration problem that Ratcliffe and other workers at the University of New South Wales [33, 134, 190, 9, 10] applied the spouted bed technique to coal carbonization, since with the use of coarse particles, the heating rate and therefore the caking tendency was decreased. Also, the high velocity spout served to break up any agglomerates as soon as they were formed. The experimental system used, in which the spouted bed carbonizer consists of a 15 cm diameter vessel, is shown in

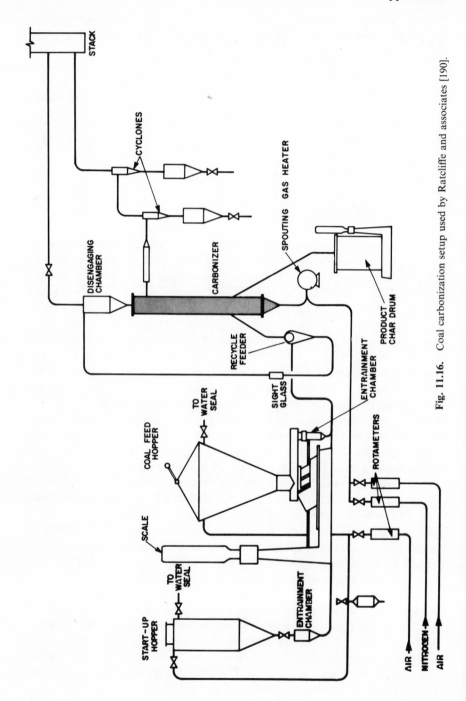

Fig. 11.16. Coal carbonization setup used by Ratcliffe and associates [190].

Fig. 11.16. Although provision was made for recycling the char if agglomeration occurred, recycling did not in fact prove to be necessary in any of the experimental runs carried out. Several different types of Australian coals were investigated, using particles of $-6 +10$ mesh size and bed temperatures in the range of 450–650°C. The heat of carbonization was provided by combustion within the bed, using a mixture of air and nitrogen as the spouting medium. The proportion of the two gases provided the means for controlling the bed temperature. Yields and analyses of the gas and char produced (but not of tar and light oil) were determined in each run.

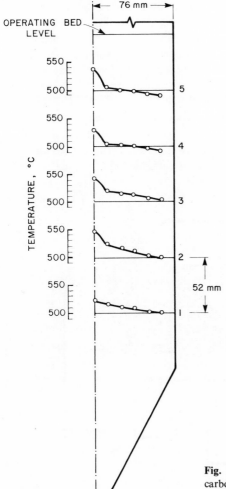

Fig. 11.17. Bed temperature profiles in the carbonizer of Fig. 11.16 measured with a bare thermocouple [190].

The rate of carry-over of fines, which contained a proportion of dust from the coal feed in addition to the char elutriated out of the bed, was found to be generally high—of the order of 15% of the coal feed. Feed rates varied between 6.4 and 10 kg/hr, the corresponding values of mean residence time being 44–29 min. The main outcome of the work was that the use of coarse particles permitted by a spouted bed, together with the attrition occuring in the high velocity spout, were effective in avoiding the problem of agglomeration for a wide variety of coals, although the oxidizing atmosphere used during carbonization must have also inhibited agglomeration to some extent.

Having demonstrated the feasibility of spouted bed carbonization, the Australian workers studied the kinetics of coal devolatilization in a specially designed thermogravimetric apparatus, under simulated conditions of gas flow and temperature determined in a carbonizing spouted bed. Temperature profiles measured with bare thermocouples showed that temperatures in the spout were considerably higher than in the annulus, while the annulus itself was substantially isothermal (see Fig. 11.17). A mean annulus temperature was used in the simulated experiments, neglecting the devolatilization in the spout in comparison with that in the annulus. The kinetic data from the simulated model, used in conjunction with particle age distribution for perfect mixing (see Chapter 4), enabled the residual volatile

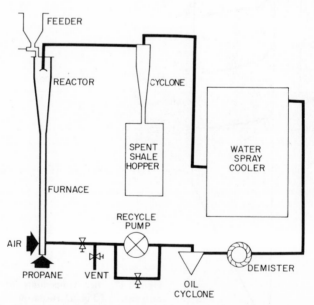

Fig. 11.18. Berti's shale pyrolysis apparatus with a tapered spouted bed reactor [23].

matter of the exit char for the spouted bed runs on Liddell coal to be predicted with a maximum deviation of 7%.

In a subsequent paper concerned with developing a kinetic model for the devolatilization reaction, Barton and Ratcliffe [10] argued that a model based on chemical kinetics alone, such as that developed by Pitt [181] for carbonization in a fluidized bed, is not applicable to coarse particles of the size used in the spouted bed process, since his assumption of particle isothermality implies that a cold feed particle would attain the bed temperature almost instantaneously. Since large particles would take a considerable time to reach the bed temperature, the devolatilization rate would depend on the particle heating rate. The argument was, however, later retracted [187] on the basis of evidence showing that the residence time for carbonization, even with coarse particles, was normally many orders of magnitude greater than the few seconds taken by the cold feed particles to attain the spouted bed temperature (see Chapter 8). The assumption of isothermality of the bed solids was therefore considered justified for predicting reaction rate from batch monolayer data for the isothermal devolatilization of coal, at a gas (nitrogen) flow rate similar to that prevailing in the spouted bed annulus. These data, for any given temperature, were found to obey the following "pseudo overall rate expression":

$$-dC_v/dt = kC_v^n \tag{11.4}$$

where C_v is the instantaneous volatile matter content of the carbonizing particles. The value of n in Eq. (11.4) proved to be independent of temperature level, while the rate constant k showed an Arrhenius type dependence with temperature. A generalized design procedure for continuous coal carbonization, which is analogous to Becker's [16] procedure for designing a drier for diffusion-controlled drying (discussed in Chapter 9), has been proposed on the basis of the above analysis, retaining the assumption of ideal mixing behavior of the spouted bed solids. The validity of this procedure is supported by experimental data on spouted bed carbonization of six different types of Australian coals at temperatures ranging between 430°C and 600°C.

B. Shale Pyrolysis

During mining and crushing of oil shale, 10–20% of the shale produced is too coarse (up to 6 mm particle size) to be effectively processed in a fluid bed retort. An investigation to determine the feasibility of carrying out pyrolysis of this coarse material in a spouted bed retort has been reported by Berti [22, 23], using the experimental system shown in Fig. 11.18. The outer surface of the particle undergoing pyrolysis becomes soft and friable due to loss of organic matter and is broken off by attrition in the spout region. Fresh surface for retorting is thus continually exposed as the particle cycles in the bed until the fine spent shale, which is also less dense than the unspent coarser material, is elutriated out of the bed. At the same time, some breakdown of cold feed particles also occurs by explosive cleavage as they enter the hot bed. The spent shale is collected in an overhead cyclone, while the gas stream proceeds through the oil recovery system consisting of a direct-contact water cooler, a demister unit for coagulating the oil mist formed in the cooler, and an oil cyclone for recovering the oil in the form of an oil–water emulsion. To maintain spouting stability, the main body of the reactor was made conical, tapering from 16.5 cm diameter at the top down to 5.1 cm diameter over a height of 45.7 cm (with a 12.7 mm diameter gas inlet), so that the large increase in volumetric gas flow along the height was offset by the increase in cross-sectional area of the bed. The removal of the reacted solids was so arranged, by locating a conical exit tube in the upper part of the reactor just above the top of the spout, that the relatively light and small spent shale particles were preferentially carried out by the gas stream, the rest of the solids falling back onto the annulus of the bed.

The process, once started, is self-sufficient in heat, the heat required for pyrolysis of the organic matter becoming available from combustion of coke which deposits on the rock, as well as by combustion of the light hydrocarbons produced. For starting up, a direct-fired propane furnace was used for preheating the bed in order to form the initial coke and to initiate its combustion, which occurs only above 480°C. The shale feed was then started and the reactor temperature controlled by adjusting the ratio of recycle gas (which is essentially free of oxygen) to air in the spouting fluid, the rate of combustion of the coke being proportional to the partial pressure of oxygen. The operational limits of the reactor were found to be 510–650°C with shale feed rates of 4.5–9.0 kg/hr. The upper limit on the feed rate was imposed not by the residence time requirement in the reactor, but by insufficient carry-over of spent shale from the reactor. Average residence times of 5–10 min are considered sufficient for complete pyrolysis, as against 15–44 min in the experimental runs. Berti therefore suggests that the capacity of the reactor could be considerably increased by discharging the spent shale (which would presumably segregate to the top of the bed) into a sealed hopper through an overflow pipe in the wall of the reactor. The oil yields, expressed as percentage of theoretical yield (hypothetical value based on total organic matter, assuming that 100% of it is recoverable as oil), were less than 50%. The low yields have been attributed by Berti to the large heat loss from the reactor wall compared to the process heat requirement ($\sim 20\%$). Higher yields would therefore be expected in a larger sized unit, where the proportion of heat lost would be substantially lower.

A mathematical model developed from the heat and mass balance around the reactor, assuming stirred-tank behavior and isothermal bed conditions, and incorporating the kinetics and thermodynamics of the reactions involved, was used to analyze the experimental results and to predict the responses to changes in the operating parameters of the system [23]. Predicted and experimental values of air requirement for combustion under different operating conditions showed close agreement, demonstrating the validity of the theoretical analysis. The model should be useful for scaling up of a spouted bed oil shale retort.

C. Iron Ore Reduction

A major difficulty encountered in fluidized bed processes for iron ore reduction is the tendency of the particles to become tacky at temperatures above 500°C, leading to buildup on the reactor walls of sponge iron and to poor fluidization. The temperature restriction thus imposed limits the reduction rate, and at least one process (H-iron process of Hydrocarbon Research Inc.) uses a high pressure, with accompanying complication in the loading and unloading operation, to raise the rate of reduction to a practical level [263, p. 23].

Laboratory scale experiments using a spouted bed reactor with both specially prepared ore–flux pellets and coarse particles of the ore itself have been carried out by Soviet workers [7, 245–247], who have reported that the special features of spouting, namely, rapid well-ordered mixing of solids, absence of dead zones and of fines from the reaction zone, and attrition of agglomerates in the spout region, enabled reduction of the ore to be carried out at temperatures as high as 700–1000°C without the problem of stickiness. Three different processes have been studied, using unusually shallow beds housed in a quartz reactor (see Fig. 11.19):

(1) Reduction of fluxed ore–fuel pellets of 2–3 mm diameter (prepared from iron ore concentrate, lime, and coke in a composition similar to that used in a blast furnace process) with water gas, to obtain sponge iron. Comparative tests carried out in a spouted bed and a packed bed showed that the process of reduction in a spouted bed began to occur intensively at 900°C, while temperatures of more than 1000°C were required in a packed bed. In spouted bed experiments, 97–98% reduction of iron in the

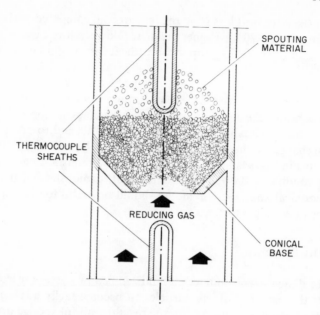

SPOUTING
MATERIAL

THERMOCOUPLE
SHEATHS

REDUCING GAS

CONICAL
BASE

Fig. 11.19. Quartz reactor used by Soviet workers for iron ore reduction experiments [7]. The reactor was surrounded by a heating furnace.

pellets was achieved in 3 min at about 1000°C without sintering the pellets or burning them onto the reactor walls [7].

(2) Reduction of the ore itself by hydrogen and by mixed gas (58% CH_4, 33.6% H_2, 6.0% CO, plus CO_2 and O_2) to obtain sponge iron and concentrate powder. Four different types of ore of size fractions 0.25–0.50 mm and 0.5–1.0 mm were tested in 20-gm batches over the temperature range 700–1000°C. With hydrogen as the spouting fluid, the degree of reduction reached 93–95% within 1 min of heating at temperatures above 700°C, while temperatures over 800°C were required with mixed gas. No difficulty due to stickiness was encountered up to a temperature of 1000°C. Detailed data on yields and chemical compositions of both sponge iron and metallic powder obtained from the different ores under various conditions have been reported [245].

(3) Reduction of ore–carbon granules of 0.5–1.0 mm and 1–2 mm size by natural gas (92.4% CH_4, 6.2% H_2, plus CO and O_2). The granules used in these experiments were made from hematite ore containing 64% Fe, with coke breeze added to give up to 15% carbon in the granules. The temperature range investigated was 800–1000°C. The temperature required for rapid reduction was found to be 900°C with low carbon content and 950°C with high carbon [247].

With the small scale of the apparatus used, it would be premature to draw any conclusions concerning the industrial feasibility of these spouted bed processes, but the results appear to be sufficiently promising to merit further investigation on a continuous pilot plant for iron ore reduction.

D. Cement Clinker Production

The insulated alumina concrete reactor portrayed in Fig. 11.20 was used recently for the experimental production of Portland cement clinker by means of a spouted bed [90]. The diameter of the gas inlet tube was 5 cm and its length 55 cm, while the reactor diameter was 15 cm. The solids fed to the reactor were dried, decarbonated cement granules of 1–3 mm diam-

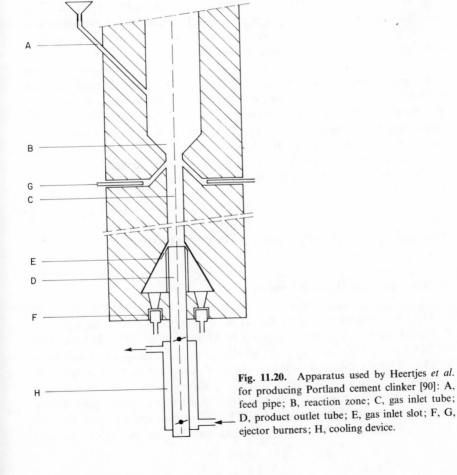

Fig. 11.20. Apparatus used by Heertjes *et al.* for producing Portland cement clinker [90]: A, feed pipe; B, reaction zone; C, gas inlet tube; D, product outlet tube; E, gas inlet slot; F, G, ejector burners; H, cooling device.

eter, while the spouting fluid was a hot mixture of air and burned natural gas. The shrinking and sticking together of the cement particles during sintering caused an increase in their terminal velocity, so that highly selective product separation could be effected in the gas inlet tube by setting the gas velocity in this tube at the terminal velocity of the clinker product. However, at the reactor temperature range of 1350–1550°C, the conversion of the feed during the particle residence time was incomplete. Further experiments are planned at 1550–1750°C as an attempt to rectify this defect [90].

E. Charcoal Activation

A recent report from Rumania [52] briefly describes the development of a spouted bed furnace for producing activated carbon, which has passed through the laboratory and pilot stages and is now in industrial use. The spouted bed was chosen in preference to a fluidized bed, first, because a coarse granular product was required, and second, to avoid operational difficulties associated with using a distributor plate at the high inlet gas temperatures involved (1000–1500°C). The spouting gas consisted of steam and combustion products obtained by burning the vapor leaving the bed with preheated air in the space above the bed. Both decolorizing and gas-absorbing varieties of active carbon in different granule sizes could be conveniently produced by varying the starting materials and controlling the operating conditions of the furnace to suit the specifications of the desired product. The criteria used for scaling-up the furnace to industrial size have been described by the Rumanian workers [51, 52].

11.6 SOLIDS AS HEAT CARRIER OR CATALYST

A. Thermal Cracking of Petroleum

The only experimental work on the use of a spouted bed for vapor phase chemical reaction has been reported by Uemaki *et al.* [237, 238, referred to in Chapter 10], who investigated the thermal cracking of petroleum— crude oil, heavy oil, and naphtha—to produce ethylene and propylene in spouted beds of coarse inert particles (alumina, chromia–alumina, and coke, in the size range 1–4 mm). High-temperature cracking of hydrocarbons (at temperatures above 600°C) is not a catalytic reaction, the role of the solid particles being to provide surface for carbon deposition, and

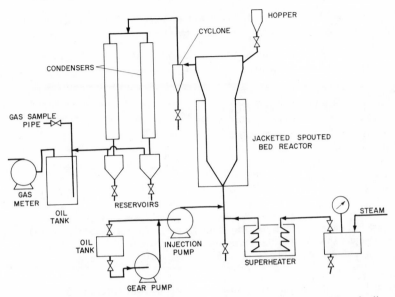

Fig. 11.21. Thermal cracking unit of Uemaki *et al.* [237]. The reactor was 13 cm in diameter with a 1.6 cm diameter gas inlet orifice.

to serve as heat carriers for the endothermic cracking reaction so that isothermal conditions are maintained in the reaction zone, namely, the bed. Since yields of ethylene and propylene are favored by a short contact time, the rationale for using a spouted bed of coarse particles in preference to a fluidized bed of fine particles was essentially the higher permissible gas flow rate in the former system.

The experimental system used (see Fig. 11.21) consisted of a 13 cm diameter × 75 cm high reactor (cone angle 60°, gas inlet diameter 16 mm), surrounded by a heating jacket for supplying supplementary heat to the bed by burning water gas. The primary source of heat supply was superheated steam, which, in combination with the oil feed pumped into the approach pipe to the reactor, served as the spouting medium. The gaseous products from the reactor, after passing through a cyclone and a condenser, were collected in a gas holder. The volume of the gas produced in each run was measured with a wet-test meter and detailed analysis of the gas samples was carried out chromatographically. An estimate of the carbon deposited on the solids was obtained from the CO and CO_2 contents of the gas, by assuming that these two gases were produced entirely by reaction between steam and the deposited carbon. A large amount of data from continuous runs, showing the effect of the operating parameters of the system (see Table 11.10 for range of conditions and materials used) on yields of total

TABLE 11.10 Materials and Conditions Used in Thermal Cracking Experiments[a]

Materials	Density (mg/m^3)	B.P. range (°C)	Particle size (mm)	Shape factor[b]	Flow rates (kg/hr)
Feedstock					0.7–3.0
Naphtha	0.690 (20°C)	30–124			
Heavy oil	0.873 (25°C)	178–360			
Crude oil (Arabia)	0.850 (20°C)	45–303			
Solids					
Aluminum oxide	2.24		0.99–2.79	0.42	
Chromia–alumina	2.90		1.65–3.96	0.45	
Coke	1.42		1.17–2.36	0.70	
Steam[c]					7–16

[a] From Uemaki *et al.* [237]; bed weight 1.0–2.5 kg; reaction temperature 550–860°C.
[b] From equation of Shirai [214] for pressure drop in fixed beds.
[c] Steam/feedstock: 4.5–12 kg/kg; nominal contact time 0.18–0.36 sec.

gas, ethylene, propylene, methane, hydrogen, and carbon for each of the three feedstocks used have been reported by Uemaki *et al.* Their main findings, as illustrated by the data for cracking of naphtha, are as follows:

(1) Yields of ethylene and propylene, as well as of total gas, increased linearly with reaction temperature up to about 800°C and began to decrease at higher temperatures. The maximum yield values obtained were 45% for ethylene, 20% for propylene, and 80% for total gas.

(2) In experiments using different bed depths at a fixed gas flow rate (and bed temperature ~820°C), the maximum yield of ethylene (~50%) was obtained at a gas–solids contact time of 0.3 sec and of propylene (~14%) at 0.15–0.2 sec. The contact times quoted are nominal values, obtained by dividing the packed height of the solids in the reactor by the superficial velocity of the spouting gas. The gas was assumed to be steam, neglecting the effect of the small proportion of oil vapors on volumetric flow rate.

(3) However, when the change in contact time was brought about by varying the gas flow rate with a fixed bed depth of solids, the dependence of ethylene and propylene yields on contact time was found to be less distinct, although yield values at the maximum contact time of 0.4 sec were noticeably lower than at shorter times. Thus a difference in behavior with respect to contact time was observed, depending upon whether the variation in contact time was achieved by varying the bed depth at a constant flow rate, or varying the flow rate at a constant bed depth. The explanation lies in the arbitrary definition of nominal contact time, which would bear

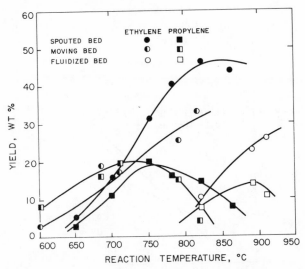

Fig. 11.22. Comparison of ethylene and propylene yields from thermal cracking of naphtha for spouted bed [237], moving bed [157], and fluidized bed [109], as reported by Uemaki *et al.* [237].

a different relationship to the real contact time in a spouted bed in the two situations, depending mainly on the distribution of gas between the spout and the annulus. Since the gas flow rate through the annulus of a given bed would remain almost unchanged with increasing total gas flow rate, the real contact time in the annular part of the bed would decrease only slightly as the gas flow rate is increased. The contact time for the spout gas would no doubt decrease considerably, but since the gas passing through the spout constitutes only a fraction of the total gas, the effective contact time in the bed as a whole, and therefore the yields of ethylene and propylene, would show only a weak dependence on the nominal contact time, as experimentally observed.

(4) A comparison of the ethylene and propylene yields for the spouted bed reactor with yields reported by other workers for fluidized and moving bed reactors, obtained under their respective optimum operating conditions, has been presented by Uemaki *et al.* (see Fig. 11.22). The maximum ethylene yield is seen to be clearly higher for the spouted bed, while the propylene yield is not significantly different. Comparative data for cracking of crude oil also showed the same trend.

Uemaki *et al.* have estimated that the operating gas velocities for the moving and fluidized bed systems of Fig. 11.22, which would be limited by hydrodynamic considerations, could not have exceeded 0.3 m/sec and 1.3

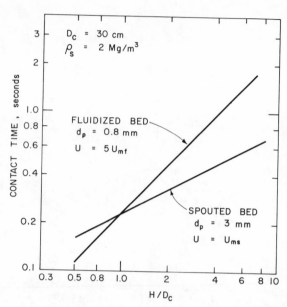

Fig. 11.23. Estimated nominal contact times in fluidized and spouted beds [237].

m/sec, respectively ($0.8U_{mf}$ for moving and $5.0U_{mf}$ for fluidized beds), as against 2.0 m/sec used by them in the spouted bed reactor. With the bed depth used in each case, which was presumably the optimum, the ratio of contact times, spouted:fluidized:moving bed works out to 0.65:1.0:4.3. The advantage of using a spouted bed is thus attributed to the short contact time achieved under optimum hydrodynamic conditions, the critical effect of contact time on the yield of ethylene in particular having already been demonstrated under item (2).

In further comparing the spouted bed with the fluidized bed in this context, Uemaki *et al.* have pointed out that for a given solid material, the only means for controlling the contact time in a fluidized bed is by reducing the depth of bed, but this cannot be done beyond a certain limit, since a sizable depth is required for maintaining isothermal conditions. The gas flow rate for spouting, on the other hand, increases with increasing bed depth, so that the contact time in this case remains short even in relatively deep beds. A comparison between the estimated contact times for a spouted bed and a fluidized bed, presented in Fig. 11.23, illustrates the above point. Thus, under typical conditions, the contact time in a fluidized bed is seen to become progressively higher than in a spouted bed, as the bed height to diameter ratio is increased above unity. A further advantage of the spouted bed over the fluidized bed mentioned by Uemaki *et al.* is the somewhat greater flexibility which the former allows in controlling the contact time, not only by changing the bed depth but also by varying the diameter of gas inlet.

Further work aimed at supplying a major part of the heat required for cracking by internal combustion of the deposited coke as in the BASF process, rather than through superheated steam, has been reported more recently [238]. A diagram of the system used, which involves operating a spouted bed combustion chamber in conjunction with the cracking unit, is shown in Fig. 11.24. The results of this investigation show that the concept is feasible and that a large saving in steam consumption can be realized by this method of operation (steam to feedstock ratio of 2–3), while maintaining high yields of ethylene and propylene.

B. Solid Catalysis

The subject of solid catalysis in spouted beds, already discussed in Chapter 10 from a theoretical standpoint, is still an open field. It is easy to imagine the technical practicability of a dual spouted bed catalytic cracker and regenerator using a TCC type pelletized catalyst [177, p. **4**.12], analogous to fluid catalytic cracking and regeneration, as well as to the

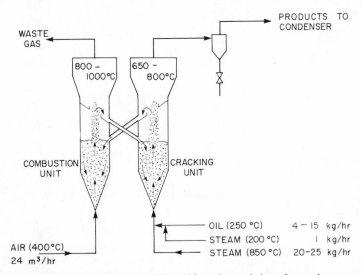

Fig. 11.24. Dual spouted bed system of Uemaki *et al.*, consisting of a cracker operating in conjunction with a combustion unit for burning of deposited carbon [238]: $D_c = 15$ cm, $H = 39$ cm, $D_i = 2.5$ cm for both units. Sand of 1.1 mm diameter was used as heat carrier; solids circulation rate was 10 kg/hr.

dual spouted bed thermal cracker and combustion chamber discussed above. Whether such a combination would be economically feasible is another question, yet to be answered. The continuing wide use of fixed packed beds of granular or pelletized catalysts and catalyst carriers as chemical process reactors, in which catalyst regeneration or reactivation requires periodic shutdown of each unit, remains an incentive for developing alternatives such as spouted beds, in which the catalyst pellets or granules can be treated continuously.

12 | *Modifications and Variations*

The spouted bed, as defined in Chapter 1 and discussed in the subsequent chapters, has been generally regarded as being associated with the simple single-spout configurations shown in Fig. 1.2. This restriction is, however, not inherent to the spouting phenomenon, which is more generally characterized by its basic hydrodynamic features rather than by a specific geometrical arrangement.

In this chapter, we discuss two alternate configurations for spouted bed operation, namely, multiple spouting and multistage spouting. A proposal to operate a spouted bed with pulsating rather than steady flow of the fluid, and suggestions to insert a physical partition between the spout and the annulus are then examined. The latter part of the chapter deals with a number of more basic modifications of the spouting technique which do not fully meet the criteria used to define a spouted bed in Chapter 1, but are nevertheless related to the spouting phenomenon in a broad sense.

12.1 MULTIPLE SPOUTING

Is there an upper limit of bed diameter beyond which a single spout would cease to cause circulation of the entire bed solids? Intuitively, it

Fig. 12.1. Rectangular sextuple-spouted bed in operation. Cross section, 41 cm × 61 cm; wheat bed, 76 cm deep. (Courtesy of W. S. Peterson, National Research Council of Canada [179].)

seems conceivable that such a limit might exist, but there are no obvious theoretical reasons which rule out the possible existence of a bed, say, 6 m in diameter by 12 m deep spouting satisfactorily with a 1 m diameter gas inlet duct. A unit of this scale has not yet been built, the largest operating spouted beds being apparently the ICI polymer blending units discussed in Chapter 11. The actual size of these blenders has not been revealed, but from the batch size figure of 2000 ft^3 (56.6 m^3) mentioned in the ICI patent [27], one would guess the bed to be up to 3 m in diameter by 6–9 m deep. Generally, two factors would weigh against single-spout beds of very large capacity.

(1) The large bed height required for effective spouting in a large diameter column ($H/D_c = 2$–3) would call for gas supply at a correspondingly high discharge pressure.

(2) The particle cycle time would be long. This may not pose a problem in an operation such as blending, but in a process involving heat transfer, e.g., drying of a heat-sensitive material, the time spent by a particle in the hot region of a very large bed could become long enough to cause thermal damage.

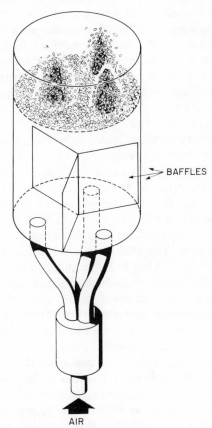

Fig. 12.2.   Circular triple-spouted bed [179].

It was mainly the latter consideration which led Peterson [179] to propose the use of multiple spouts within the same bed as a means for maintaining the required rapid circulation of particles for large capacity drying. In preliminary experiments to develop a multispout system, Peterson found that although several individual spouting cells, each with its own spout and annulus, could be established in a large bed by choosing suitable spacing of gas inlet openings, operation tended to be unstable except with shallow beds. Provision of vertical baffles, extending upward from the base of the vessel and covering one-half to seven-eighths of the height of the bed, brought about a marked improvement in multispouting stability, presumably by cutting off lateral flow of gas between neighboring cells. Exchange of solids between cells was still permitted through the upper unpartitioned section of the bed. A photograph of an operating sextuple-spouter, designed on the above principle, is shown in Fig. 12.1. The unit shown could be operated continuously, with feed and overflow points at opposite ends, without any noticeable effect on spouting behavior (maximum feed rate tried—590 kg/hr; mean residence time ~12 min). Also, solids mixing in the bed at this feed rate, as determined by introducing a slug of tracer particles with the feed, was found to be nearly perfect, as in a single-spout bed. Hence, for large-scale operation, a multiple-spout bed of the type shown in Fig. 12.1 would appear to be a feasible alternative to using either more than one or a larger single-spout system.

Based on his experimental results from the sextuple-spouter of Fig. 12.1, as well as those from a circular (Fig. 12.2) and a rectangular triple-spout configuration, Peterson concluded that the minimum air flow required in multiple spouting, with the air-tight baffles which he used between cells, is roughly 20% higher than the sum of the air flows necessary for spouting of each single-spout cell as given by Eq. (2.38). The latter flow rate for the noncircular cells was the same as for spouting in a round column of equal cross-sectional area. The wall particle velocities, on the other hand, were found to be nearly identical to those obtained in a round column with the same cross-sectional area as one cell of the multiple-spout bed.

A comparison of the observed performance of the sextuple-spouter of Fig. 12.1 with the estimated properties of two hypothetical single-spout beds, one having the same bed volume as the sextuple unit and the other the same air flow rate, is shown in Table 12.1. It is seen that the multiple-spout system provides more vigorous agitation of a given mass of solids than the equivalent single-spout bed, but at the expense of higher gas consumption; also, it allows a shorter residence time for the gas. To achieve the same flow rate of gas as in the sextuple unit, the single-spout bed required is much larger, provides slower solids turnover and longer gas residence time, and requires gas supply at a higher pressure. The choice

TABLE 12.1 Comparison between Multiple-Spout and Single-Spout Beds[a]

	Sextuple-spouter of Fig. 12.1 (Peterson's observed data)	Estimated values for equivalent single-spout beds	
		Same bed volume	Same volumetric air flow rate
Size			
Cross section	41 cm × 61 cm	46 cm diameter	76 cm diameter
Cross-sectional area (m^2)	0.25	0.16	0.46
Bed depth H (cm)	76	107	190
Bed volume (m^3)	~0.14	0.14	0.68
Minimum spouting air flow			
Volumetric (m^3/sec)	0.219	0.097	0.219
Superficial velocity U_{ms} (m/sec)	0.88	0.58	1.2
Average gas residence time			
H/U_{ms} (sec)	0.9	1.8	1.6
Solids circulation			
Particle velocity at wall (m/sec)	27.9	17.8	12.7
Particle cycle time (sec)	22	60	150
Solids mass flow (kg/sec)	5.44	2.28	4.53
Bed pressure drop (kN/m^2) $= \frac{2}{3}\dfrac{\text{bed weight}}{\text{cross-sectional area}}$	4.2	5.8	10.4

[a] System: air–wheat.

between a multispout and a single-spout system would therefore depend on the requirements of the particular process with respect to solids turnover rate and gas residence time on the one hand, and the economics of gas supply on the other. For example, a multispout bed would be preferred for drying a particularly heat-sensitive material where rapid turnover is of critical importance. For granulation, on the other hand, a single-spout system is likely to be more suitable, since it would allow more time for drying of a granule between deposition of successive liquid layers.

The operational stability of a multispout system is more delicate than that of a single-spout bed, being vulnerable to imbalance between the gas flows to individual cells as well as to interference between neighboring cells in the upper part of the bed. The gas supply must therefore be so arranged that a constant flow ratio between the cells is maintained over a range of gas flows to facilitate start-up, and subsequently the spout heights in all the cells should remain equal. With a common manifold for air distribution, the preceding requirements could be met in Peterson's experiments only if the resistance to air flow in each inlet line were kept similar by using identical lengths of pipe, and if this resistance were high enough to give a back pressure at the manifold equal to about three times the pressure drop across the bed itself. Peterson therefore recommends the use of individual flow controllers on the gas supply lines, for economy in compressor costs. He also recommends the use of conical spout deflectors, which can be seen in Fig. 12.1, in order to minimize interference between neighboring cells caused by spout wandering as well as to prevent blowout of solids during start-up.

While the stability of a multispout bed is very much improved by cutting off lateral flow of gas between cells through the use of partition walls, as demonstrated by Peterson, the same solids movement pattern can nevertheless be achieved without partitions if the bed is shallow. Such a system, namely, one with a shallow bed and no partitions, is in industrial use for cooling of fertilizers and has been described in Chapter 11.

12.2 MULTISTAGE SPOUTING

Two different modes of multistage operation, which involves flow of solids in series through more than one spouted bed, are possible: (a) with the flow of spouting gas also in series, and (b) with the flow of spouting gas in parallel. Some experimental work with the former system only has been reported, by Madonna et al. [125], who demonstrated that up to four beds of wheat located one above the other in a 15 cm diameter column could be satisfactorily spouted with the same air. Malek and Walsh [130] subsequently tried such a system for continuous heating of solids (sand and coal) which traveled downward from one bed to the next through internal downcomers (see Fig. 11.11) countercurrent to the flow of hot spouting gas. They also used a 15 cm diameter column with up to four stages.

The important features of continuous operation in a multistage system of this type are:

(1) Since spouting of successive beds is achieved by the same gas, the gas flow rate required for spouting a given aggregate depth of solids is less than that for a single bed of the same depth. In a system of N stages with an aggregate bed depth H, the gas flow required for spouting each bed (of depth H/N) will, according to Eq. (2.38), obey the following relationship:

$$[Q_H]_{\text{multistage}} = K(H/N)^{1/2} \tag{12.1}$$

while the flow rate for spouting a single bed of depth H will be

$$[Q_H]_{\text{single bed}} = KH^{1/2} \tag{12.2}$$

Therefore,

$$[Q_H]_{\text{multistage}} = \frac{[Q_H]_{\text{single bed}}}{N^{1/2}} \qquad (12.3)$$

Experimental results of Madonna *et al.* for wheat are in approximate agreement with this relationship. For example, the observed air flow rates for spouting a total depth of 30 cm in a two-equal-stage unit and in a single bed were 40 m^3/hr and 51 m^3/hr, respectively, the ratio between the two being 0.78 as against 0.71 predicted by Eq. (12.3).

(2) The economy in gas supply is, however, achieved at the expense of back pressure, since pressure drop occurs due to contraction of gas at the entrance to each stage, over and above the pressure drop across each bed itself.

(3) The lower gas velocity in a multistage unit would permit a longer gas contact time with the solids, compared to that in a single bed of the same height as the combined beds of the multistage unit. This would be beneficial for certain processes; for example, the thermal efficiency in heating or cooling of solids would be improved by multistage operation.

(4) A more important benefit would arise from narrowing down of the solids residence time distribution. The wide distribution in a single spouted bed, which is inevitably associated with its nearly perfect solids mixing characteristics, can be a serious disadvantage if uniform solids treatment is desired. It has been shown in Chapter 4 (and Chapter 9) that the residence time or exit age distribution in a continuously operating spouted bed is quite well described by the following perfect-mixing equation:

$$E(\theta) = e^{-\theta} \qquad (9.18b)$$

which is equivalent to

$$E(t) = \frac{1}{\bar{t}} e^{-t/\bar{t}} \qquad (12.4)$$

where $E(t)\,dt$ is the fraction of solids staying in the bed for the time interval between t and $t + dt$, and $\bar{t}$ is the mean residence time. The same equation applies to a fluidized bed [110, p. 327]. Therefore, the exit age distribution in a multistage spouted system of N equal-sized beds should also be described by the multistage fluidized bed equation given by Kunii and Levenspiel [110, p. 329]:

$$E(t) = (1/(N-1)!\bar{t}_i)(t/\bar{t}_i)^{N-1} \exp(-t/\bar{t}_i) \qquad (12.5)$$

$\bar{t}_i$ being the mean residence time in each stage. Equation (12.5) reduces to Eq. (12.4) for $N = 1$. The exit age distributions for solids in single and

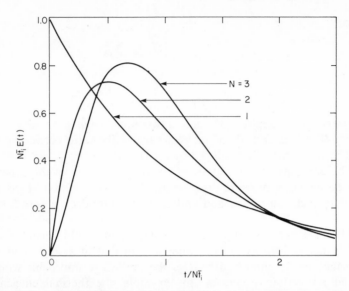

Fig. 12.3. Exit age distributions of solids for multistage systems with perfect solids mixing in each stage, calculated by Eq. (12.5) [110, p. 329].

multistage beds, as given by Eq. (12.5), are shown in Fig. 12.3, which is reproduced from the book by Kunii and Levenspiel.

It should be noted that the increase in uniformity of solids residence time with number of stages shown in Fig. 12.3 will be realized, regardless of whether the spouting gas flows in series through the different stages, as in Fig. 11.11, or in parallel. The choice between these two configurations for carrying out a particular process in a multistage system would therefore depend on the other considerations discussed, such as, gas flow rate, pressure drop, and gas residence time.

12.3 PULSED FLOW SPOUTING

From the discussion in Chapter 10, one general shortcoming of the spouted bed as a vapor phase chemical reactor is obvious: the unequal contact with solids of the gas passing through the dilute phase spout and the dense phase annulus, which could have a detrimental effect on yield and selectivity for many chemical reactions. Pulsating the flow of spouting gas has been proposed as a means for minimizing the nonhomogeneity of gas treatment while maintaining the spouting pattern of solids circulation [249, 251, 56].

The general effect of alternately introducing the gas at a high enough velocity to initiate spouting and then cutting off the gas supply is to bring about a reduction of the time-averaged voids in the spout region, since particles separated by the frictional drag of the gas during the active phase of the cycle would tend to revert back to their original packed bed orientation during the inactive period. The exact behavior of a given bed under pulsed flow conditions would depend on the number of cycles per unit time (frequency), fraction of the cycle time over which there is no gas flow (intermittency), and the rate of gas flow during the active period (amplitude). For example, with a sufficiently long duration of each period, that is, with low frequency together with high intermittency, the bed can be made to alternately spout and become packed. Under these conditions, the time-averaged spout voids would be smaller than if the bed were to spout continuously, but the time-averaged gas flow rate would also decrease. If, on the other hand, the active phase of the cycle is so brief (i.e., high frequency and high intermittency) that the gas flow is cut off while the spout is still developing, then the spout will never pierce the bed even with a high amplitude, and yet over a period, high gas flow rates and particle circulation rates approaching those in a continuously spouting bed could be achieved. This is illustrated in Fig. 12.4, which shows the movement of particles and development of the spout in a two-dimensional bed based on observations made by Volpicelli *et al.* [251] using a cine camera. The left-hand side of the diagram represents the active phase of the cycle and the right-hand

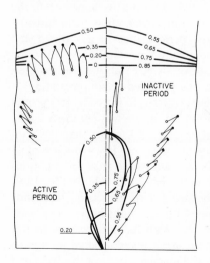

Fig. 12.4. Changes in bed structure and particle movement during one cycle in pulsed flow spouting [251]. Column cross section, 20 cm × 1.5 cm; $H = 11$ cm; $d_p = 3.1$ mm; air inlet, 4 mm wide slit extending across the full 1.5 cm width of the base; air flow during active phase = 0.5 gm/sec; pulsation frequency = 3.2 Hz; intermittency = $\frac{1}{2}$; ○ position of tracer particles at the start of the active period; ● position of tracer particles at the start of the inactive period. Numerical values shown represent time, expressed as fraction of the total cycle time starting from the beginning of the active period. Heavy lines show the configuration of the cavity and of the bed surface.

side the inactive phase. It is seen that the internal spout or cavity formed during the active period became detached from the inlet orifice, shrank, and moved upward during the inactive period, the region below the cavity becoming filled up by surrounding particles sliding into it as the cavity traveled upward. At the same time, the surface of the bed bulged upward during the active period, reverting back to its original position at the end of the inactive period. Repetition of this phenomenon, in successive cycles, induced an overall pattern of particle circulation similar to that in a continuously spouting bed, but with a superimposed jiggling movement, also shown in Fig. 12.4. If, however, the frequency of pulsation is so high that the particles do not get sufficient time to slide into the axial region during the inactive period, the spouting pattern of circulation would not be achieved.

In parallel experiments with fluidized beds fed by a pulsating flow of gas, Volpicelli *et al.* found that although alternate expansion and contraction of the bed with consequent improvement in contacting effectiveness could be achieved in a fluidized bed too, within certain limits of operating conditions, no lateral displacement of solids occurred, the particle movement being only up and down in the same vertical plane. The authors have pointed out that the inward lateral flow obtained in a pulsating spouted bed is particularly advantageous from the point of view of wall-to-bed heat transfer; the benefit would in fact be more general in terms of better solids intermixing than in pulsed flow fluidization.

The average pressure drop across the bed with pulsed flow operation was considerably higher than with steady gas flow in both spouting and fluidization experiments, and this was interpreted to indicate that gas–solids interaction is more effective in pulsating beds. In order to demonstrate the favorable effect of pulsating flow on contacting effectiveness more directly, Volpicelli [249] determined the catalytic performance of a spouted bed reactor operating under pulsed flow and steady flow conditions, choosing the decomposition of ozone to oxygen on iron oxide catalyst as the reaction for the comparative tests. The catalyst was prepared by impregnation of porcelain chips of 20–25 mesh size with ferric nitrate, followed by its thermal dissociation until evolution of nitrogen dioxide vapors was completed. The results from this investigation, reported in Fig. 12.5, show that the conversion of ozone obtained at low frequencies (< 1.6 Hz), which simply caused intermittent spouting of the bed, was similar to that for a continuously spouting bed (horizontal lines on the left-hand side of Fig. 12.5). A steady improvement in conversion occurred with increasing frequency until packed bed values (horizontal lines on the right-hand side) were approached at frequencies of around 4 Hz. With further increase in frequency, however, the conversion either remained unchanged or started to decrease, suggesting that the duration of the inactive phase of the cycle had

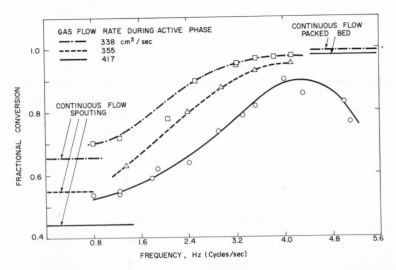

Fig. 12.5. Effect of gas pulsation frequency on degree of ozone conversion [249]. Column cross section, 18.6 cm × 0.8 cm; $H_0 = 10$ cm; bed weight = 135 gm; d_p, 20–25 mesh. Gas inlet, 3 mm wide slit extending the full 0.8 cm width of the base. (Corrected gas flow rates provided by Professor Volpicelli.)

become too short to allow settling of the particles, and therefore the voidage distribution in the spout was once again approaching that of a continuously spouting bed. Thus, the favorable effect of pulsating the gas flow on contacting effectiveness, predicted by Volpicelli *et al.* from their hydrodynamic study of a spouted bed, was confirmed by its performance as a chemical reactor.

An attempt to quantitatively relate the ozone conversion data with bed hydrodynamics was also made by Volpicelli [249], who numerically integrated the differential equations describing the flow distribution for successive conditions of the bed during a pulsating cycle in conjunction with the reaction rate equation, the former being based on his earlier mathematical analysis pertaining to the onset of spouting, outlined in Chapter 2. The trend of conversion with pulsation frequency predicted in this manner agreed with the experimental trend, but the calculated values were consistently higher, possibly because the assumption that the packed bed surrounding a cavity behaves isotropically throughout the pulsation cycle, which had to be made for calculating the flow distribution, is not strictly valid.

Additional information on the structure of a pulsating spouted bed is provided by the work of Elperin *et al.* [56], who were able to determine the time-averaged voids at various positions in a two-dimensional conical bed

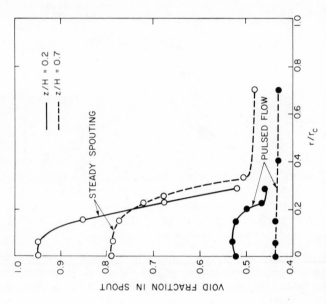

Fig. 12.7. Radial voidage profiles in steady spouting and pulsed flow spouting [56]. Same system as for Fig. 12.6.

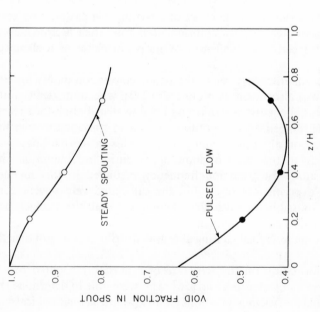

Fig. 12.6. Longitudinal spout voidage profiles in steady spouting and pulsed flow spouting [56]. Two-dimensional conical vessel, 1.2 cm wide; $H = 10$ cm; gas inlet, 1.2 cm × 1.2 cm; $\theta = 40°$. Voidage figures are time-average values, recorded over a 10 sec period for steady spouting and over 100 sec with pulsed flow.

by measuring the absorption of β rays passing as a directed 3 mm diameter beam through the bed. The solid materials studied were millet ($d_p = 2.2$ mm) and poppy seed ($d_p = 1.0$ mm), with pulsation frequencies in the range 2–16 Hz and average gas flow rates up to twice that for minimum spouting; intermittency values for these tests have not been reported. Some of the experimental results reported by Elperin *et al.* have been redrawn as Figs. 12.6 and 12.7 to illustrate the startling effect of pulsed flow on spout voidage, which is seen to decrease nearly to the packed bed value at all levels (Fig. 12.6). The radial voidage profile in the spout (Fig. 12.7) became flatter, and even the annular solids packed somewhat more tightly than in steady spouting. Visual observation showed that the usual sharp division between the spout and the annulus disappeared in the pulsating bed at frequencies above 2 Hz, as would be expected from the voidage data, but the pattern of solids movement was still preserved. The optimum gas velocity for pulsed flow spouting was found to be just above that for minimum spouting, since the decrease in spout voidage at higher gas velocities was less pronounced than shown in Fig. 12.6.

As a follow-up of the work described above, Elperin *et al.* investigated the effect of gas pulsations on what they call gas-to-particle heat transfer, but is really combined heat and mass transfer, since the moisture content of the bed solids after spouting with warm air (35–40°C) for periods of 90–150 sec is reported to have decreased from 25–32% (initial) to 22–28% (final). A conical-cylindrical column of 4.8 cm diameter with a 1.2 cm air inlet and a 40° cone was used in these experiments, the depth of bed being 6 cm. The heat transfer coefficient measured under these conditions was found to increase to a maximum value at frequencies in the region of 2 Hz and then to remain substantially unchanged (up to 12 Hz). The maximum increase over the steady spouting coefficient occurred at relatively low gas velocities ($1.15 U_{ms}$) and amounted to 15% for millet and 40% for poppy seed. The greater increase in the case of poppy seed was attributed to the fact that it tended to stick together during steady spouting (unlike millet, which was free flowing) but not under pulsed flow conditions. The authors therefore suggest that pulsed flow operation should be particularly advantageous for materials which are not completely free flowing.

An altogether different type of pulsating spouted bed, with a continuous spout but a pulsating annulus, has been described by Golubkovich and associates [81, 82]. The pulsating movement causes particles to detach themselves from the lower part of the bed and to flow out of the spouting vessel in spurts through the central opening at the base, against the current of incoming gas. Solids are fed in continuously at the top so that a "pulsating spouting bed" with a net downward movement of solids is established. The hydrodynamics of such beds has been studied in considerable detail by Golubkovich *et al.* and the critical parameters for achieving the pulsating movement as a steady condition have been identified, but no specific applications of this unusual system are mentioned.

12.4 SPOUTING WITH TUBULAR INSERTS

The use of a vertical pipe in the axial region of the bed as a start-up procedure has already been discussed in the previous chapter (Fig. 11.13). A tubular insert can also be used in steady operation to modify the normal flow patterns, as long as a sufficiently long lower section of the spout is left unenclosed to allow recirculation of solids. Cross flow of both gas and solids between the enclosed section of the spout and the surrounding annulus can be eliminated if the insert is impervious to both phases (e.g., an ordinary pipe), while only solids cross flow will be avoided by the use of a tubular screen insert. Both schemes have been experimentally investigated.

A. Pipe Insert

The "fluid-lift solids recirculator" developed by Buchanan and Wilson [34] is based on the idea that as far as solids recirculation is concerned, flow of fluid through the annulus region of a normal spouted bed serves little purpose. The use of a pipe insert or "draft tube," as shown in Fig. 12.8, forced most of the gas to travel up through the axial region without the opportunity for flaring out into the annulus, except over the short distance between the fluid inlet and the lower end of the tube. The gas flow rate required for achieving solids recirculation in a given bed was therefore reduced by the presence of the draft tube. The experimental results obtained by these workers in a 15 cm diameter wheat bed (see Table 12.2), however, show that any reduction in gas flow rate was accompanied by a corresponding decrease in solids recirculation rate. Thus, by reducing the draft tube separation from 18.8 to 3.3 cm, the minimum air flow required for spouting decreased by a factor of two, but this was accompanied by a sevenfold decrease in solids circulation rate, causing the ratio of grams solids circulated per gram of air to become smaller with decreasing draft tube separation. With the maximum separation of 18.8 cm, both air flow and solids flow rates closely approach the values reported by Mathur and Gishler [137] for a similar bed with no draft tube. Hence, while operation without a draft tube offers the most favorable conditions for solids mixing,

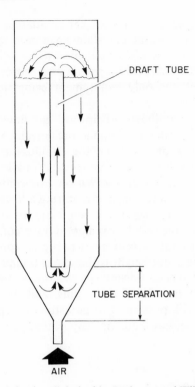

DRAFT TUBE

TUBE SEPARATION

AIR

Fig. 12.8. "Fluid-lift solids recirculator" devised by Buchanan and Wilson [34].

TABLE 12.2 Solids Recirculation Data for a Spouted Bed with a Central Draft Tube[a]

Draft tube separation (cm)	Air flow rate (gm/sec)	Solids circulation rate[b] (gm/sec)	gm solids circulated gm air flow
3.3	7.56	55.9	7.4
3.8	7.78	62.75	8.0
4.3	8.39	82.39	9.8
7.4	9.83	111.89	11.4
9.9	11.72	244.94	20.9
11.7	12.33	266.86	21.6
13.7	.13.97	307.69	22.0
18.8	15.11	405.19	26.8

[a] See Fig. 12.8; system: air–wheat, $D_c = 14.6$ cm, $D_i = 12.7$ mm; size of draft tube: 3.8 cm inner diameter by 30 cm long.
[b] Calculated from particle velocity at wall data.

the use of a tube enables weaker circulation to be achieved by a reduced expenditure of gas, thereby allowing greater operational flexibility. Furthermore, it permits solids recirculation to be achieved without the limitations with respect to particle properties and bed depth which apply in a normal spouted bed, the only requirement being that the solids should be free flowing.

With the curtailment of upward gas flow through the annulus by the draft tube (the eductor action at the bottom is reported to have even caused a reverse flow of gas down the annulus under certain conditions), it is difficult to see how such a system can provide good gas–solid contacting, as claimed by Buchanan and Wilson. For solids blending, on the other hand, it does offer the advantage of requiring a lower minimum gas flow rate and pressure drop compared to a bed without a draft tube. The corresponding reduction in the solids circulation rate combined with the absence of mixing by lateral movement of particles from the annulus into the spout all along the spout height is, however, bound to reduce the overall mixing efficiency. Other disadvantages mentioned by the authors are the risk of plugging of the tube by the solids, and the complication of installing a draft tube with provision for adjusting its separation distance as an operating variable. No industrial application of such a system has so far been reported.

B. Tubular Screen Insert

Pallai and Németh [174, 175] reasoned that spouted bed solids drying, which occurs mainly in the annulus region, would be more efficient if particles were forced to spend the maximum possible time in the annulus during each cycle without the freedom to short-circuit into the spout except in the lowermost part of the bed. Such a modification of the solids flow pattern would also improve gas-to-particle heat transfer, since all the bed particles would then pass through the hot lowermost region of the bed during each cycle. But in contrast with the solids recirculator discussed above, any curtailment of annular gas flow in a drier is obviously not acceptable from the point of view of drying efficiency.

The above requirements appear to have been met by the use of a metallic screen insert of the same shape as the spout, with its lower end kept sufficiently far away from the gas inlet orifice to avoid interference with overall solids circulation. Pulse tracer experiments, with solids fed continuously into the upper part of the annulus and discharged from the bottom, showed the residence time distribution to be narrower with the insert than without it. The insert also caused the average residence time to decrease by about

15%, indicating increased bypassing. This effect has been attributed by Pallai and Németh to the greater probability of annulus particles reaching the bottom discharge point when cross flow into the spout is absent at higher bed levels than when it is present. If this explanation is correct, it should be possible to reduce bypassing by discharging the solids from the top through an overflow pipe as in Fig. 11.1, instead of from the bottom.

12.5 DERIVATIVE TECHNIQUES

A. Ring Spouted Bed

The ring refers to the shape of the gas inlet, the bed of solids being in the annular space above, between the wall of the main vessel and an overflow weir located in the middle of the vessel. By directing the supply of gas upward through the bottom of the annular space, Romankov and coworkers [201, p. 129; 199, p. 587] were able to establish normal spouting of solids, and used the arrangement shown in Fig. 12.9 successfully for continuous drying of pastes and solutions on a pilot plant scale. The solids discharge centrally through the weir, which can be raised or lowered to

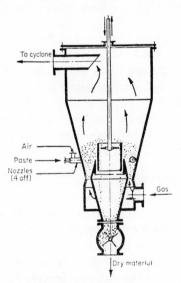

Fig. 12.9. Ring spouted bed drier for pasty materials, developed by Romankov and coworkers. (Reproduced from Romankov [199, p. 587].)

regulate bed height. This configuration is recommended by the Leningrad workers for large-scale operation, since it permits the spouting bed volume to be increased without a corresponding increase in bed height and therefore in pressure drop across the bed, as would be necessary for scaling up a normal spouted bed.

B. Slot Spouted Bed

Another Leningrad innovation [201, p. 10] involves introducing the gas into the base of the bed through a narrow slot running across the longer dimension of the containing vessel, which is of rectangular cross section with sloping walls, as in Fig. 12.10. The slot may be either continuous or with spaced openings; in either case, stable spouting of solids can be achieved within certain limits of the usual spouted bed parameters, namely, vessel size, bed depth, slot width, gas flow rate, and solids properties. The effect of these parameters on bed behavior has been investigated by Mitev [156], and gas flow oscillations which occur during spouting in such an apparatus have been studied by Volkov et al. [248].

If the slot inlet is so arranged that the spouting gas enters the bed tangentially, as shown in Fig. 12.11, rather than vertically as in Fig. 12.10, the bed solids are transported upward as a dilute phase along one wall of the column and slide downward as a dense phase along the opposite wall. The vortex movement thus established is reported, again by Mitev, to be more stable than with vertical introduction of the spouting gas, and could be achieved for a wide variety of solids, including polydispersed materials. Operation with tangential gas entry in a multistage unit, as well as in conjunction with the internal weir arrangement of Fig. 12.9, was also shown to be feasible. The successful development of industrial scale equipment based on the tangential entry principle for dehydration of gypsum, as well as for granulation of KCl and $C_6H_5(SO_3)_2Na$, was reported by Rashkovskaya at the 1972 Prague Conference [189].

Circumferential entry of gas through a narrow slot to achieve solids circulation has also been used by workers at the United States Department of Agriculture [198] in developing the hot air grain popper shown in Fig. 12.12. The air jet blows the grain to the top of the chamber. The grain then rains down onto the sloping floor of the chamber and slides along the floor to the circumference, to be reentrained in the air stream. The chamber is divided into compartments, and revolves to allow continuous feeding and discharging of material. Since the popping process requires a very short contact time with the hot air (15–30 sec), a bed of solids is not allowed to build up in the chamber, as it does in Mitev's system.

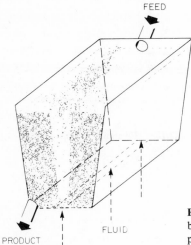

Fig. 12.10. Schematic diagram of a slot spouted bed. (After Romankov and Rashkovskaya [201, p. 10].)

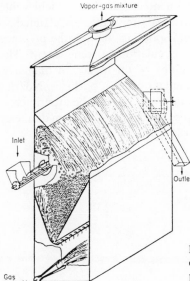

Fig. 12.11. Slot spouted bed drier with tangential entry of gas. (Reproduced from Romankov [199, p. 581].)

C. Spouting with Reverse Gas Flow in Annulus

The system shown in Fig. 12.13 has been proposed by Ageyev *et al.* [2] as being particularly suitable for carrying out catalytic processes. With the vessel top sealed, the entire spouting gas is forced downward through the

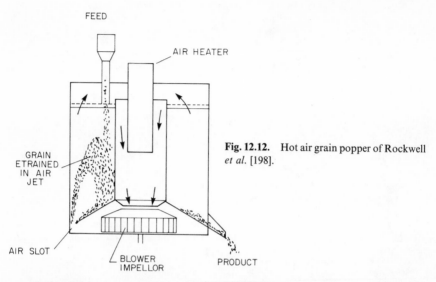

Fig. 12.12.   Hot air grain popper of Rockwell *et al.* [198].

annular solids, giving more uniform treatment of gas than is achievable in a normal spouted bed. The pattern of solids movement is claimed to remain substantially the same as in normal spouting, though the presence of the horizontal grid is likely to impede circulation to some extent. The system described has been tested in columns of up to 20 cm diameter, using 6–14 cm deep beds of 1–2.5 mm diameter catalyst particles. Nozzles of different diameters, in the range 6–28 mm, were used in the experiments. A multispout system based on this principle, consisting of a 20 cm diameter column with 30 jets spaced 3 cm apart, was found to work well.

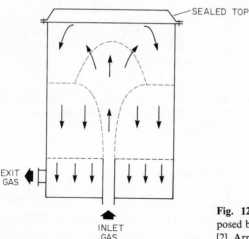

Fig. 12.13.   Top-sealed spouted bed proposed by Ageyev *et al.* for catalytic processes [2]. Arrows indicate direction of gas flow.

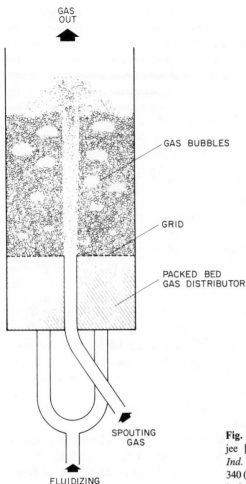

GAS
OUT

GAS BUBBLES

GRID

PACKED BED
GAS DISTRIBUTOR

SPOUTING
GAS

FLUIDIZING
GAS

Fig. 12.14. Spout–fluid bed. (After Chatterjee [37].) Reprinted with permission from *Ind. Eng. Chem. Process Des. Develop.* **9**, 340 (1970). Copyright by the American Chemical Society.

D. Spout–Fluid Bed

The idea of superimposing spouting action on an already fluidized bed, or vice versa, has received the attention of several investigators (see Fig. 12.14).

For granulation of ammonium sulfate, Romankov *et al.* [201, p. 228] created zones of local spouting with very hot furnace gas (900–1000°C) in a bed of seed granules gently fluidized with a separate air supply of lower temperature (up to 146°C), which was kept below the decomposition temperature of ammonium sulfate. The spouting gas was introduced via nozzles passing either through the fluidizing air distributor which formed the base of the bed or through the column wall at a level just above the distributor, while injection of feed solution was restricted to the high-temperature spout zones in the bed. This arrangement enabled a five- to sixfold increase in water evaporation rates to be achieved, compared to those obtained in an ordinary fluidized bed of the same size using 180°C air. Also, a more uniform size product, which did not tend to cake up during storage, was obtained. A small industrial plant, based on the spout–fluid principle described, with a capacity of 6000 kg of granulated ammonium sulfate per 24 hr, is in operation in the USSR. A commercial unit, based on essentially the same idea, has also been developed by the Nautamix Company of Holland, and is being marketed under the trade name Vometec Fluid Bed Reactor [163]. This unit is claimed to be suitable not only for granulation but also for a wide variety of other processes requiring intimate contact between solid particles and a fluid—either gas or liquid, including chemical reaction.

Two investigations into the hydrodynamic behavior of a bed in which spouting and fluidization occur simultaneously have been reported [37, 182]. According to Chatterjee [37], such a system, like ordinary spouting, overcomes the limitations of stratification and slugging inherent in a fluidized bed, but without the restrictions with respect to particle size and bed depth which are associated with the stability of the usual spouted bed. The total gas flow required for "spout-fluidizing" a given bed exceeds that for either spouting it or fluidizing it, but Chatterjee argues that the additional gas is well spent, since with the annulus fluidized, the effectiveness of gas–solid contact is improved and so is the intermixing of solids. The latter point in particular gains support from the independent observations of Pomortseva and Baskakov [182] made in a semicircular column, where bubbles from the fluidized region were seen to rush laterally toward the center and merge with the spout. The solids movement thereby induced contributed to making overall intermixing in the bed more intensive than in either fluidization or spouting. These workers extended their investigation of a spout–fluid bed by carrying out detailed measurements of submerged object-to-bed heat transfer [182].

A recent development in this area is a spout–fluid bed operating with a liquid rather than a gas as the fluid medium. Preliminary work on a liquid phase spout–fluid bed has been carried out by Vuković et al. [253], who studied pressure drop characteristics and fluid flow requirements using a bed of 1.2 mm glass beads with water as the fluid medium.

E. Two-Fluid-Phase Spouting

While the use of an immiscible fluid jet to agitate a liquid or slurry is a common procedure, carrying out a similar operation in such a way as to produce a characteristic spouting circulation pattern involves more circumscribed design limits. The operations described below were inspired by fluid–solid spouting and have wider objectives than simple agitation.

1. *Liquid–Liquid Spouting*

A spouted mixer–settler for liquid–liquid extraction has been developed by Johnston, Robinson, and Epstein [98]. The heavy liquid phase is spouted upward through a depth of light liquid phase; after reaching a certain height, it breaks up into a shower of discrete droplets which then settle through the light phase in the annular region. Very effective contact between the two liquid phases is thus obtained, without the necessity of expensive mechanical agitation. Murphree spouted-phase efficiencies of up to 96% were recorded in experiments to extract benzoic acid from toluene using water as the spouting liquid. Mixing and settling could thus be carried out continuously in a single vessel. Although the system described was suggested by and named after fluid–solid spouting, the resemblance between the two is only superficial, as pointed out by Johnston *et al.*

The mixing behavior of a jet of single-phase liquid introduced through the base of a spouting type vessel ("spouted liquid reactor") has been investigated [87, 226] and might have some bearing on liquid–liquid spouting.

2. *Gas–Liquid Spouting*

Analysis of the working of his spouted bed granulator (discussed in the previous chapter) led Berquin [21] to suggest that the spouting phenomenon could be usefully adapted for operations requiring effective gas–liquid contact, employing essentially the same apparatus as for gas–solid spouting. Experimental work on concentrating phosphoric acid from 30% P_2O_5 to 55% P_2O_5, with feed acid continuously atomized into the hot (521°C) air stream, showed that internal recycling of the liquid "bed" was induced by the air jet, giving the benefits of a well-stirred contactor. The temperature of the exit vapors (95°C) closely approached the exit acid temperature (92°C), and consequently a high thermal efficiency (75%) was achieved. The energy consumption, with air requirement at 10 Nm^3/kg concentrated P_2O_5, was also considered to be competitive with other concentration systems. The simplicity of the apparatus is mentioned by Berquin as one of the main attractions of the process, especially since with rapid cooling of the entering air, relatively common and therefore inexpensive materials of construction can be used. Operational convenience and flexibility are quoted as other important advantages; thus even superphosphoric acid of up to 75% P_2O_5 strength could be produced in the same apparatus by adjusting the operating conditions. Berquin considers that the technique of gas–liquid spouting should be well suited for concentrating other liquids too, e.g., various acids and salt solutions, as an alternative to the usual vacuum evaporation processes.

The potential of gas–liquid spouting has also been recognized by Németh and Pallai [166], who envisage a wide area of possible applications, including absorption, crystallization, and chemical reaction.

3. *Three-Phase Spouting*

It has been found possible to establish a three-phase spouted bed, using the gas phase as the spouting medium with the liquid flowing downward through the bed. The first report on the behavior of such a system has been presented recently by Vuković *et al.* [254]. Using air, water, and light spheres of polyethylene and polystyrene ($d_p = 1$–2 mm, $\rho_s = 0.23$–0.32 Mg/m^3) in a 19.4 cm diameter column, these workers determined the conditions necessary for maintaining a stable three-phase spouted bed and studied its pressure drop characteristics. They consider that the regular cyclic movement of particles combined with true countercurrent contact between the two fluid phases should make the spouted bed a more effective three-phase contactor than a fluidized bed for certain applications. Removal of dust particles from a gas stream by scrubbing and reactions involving precipitation are cited as specific examples. Application to leaching processes is another possibility suggested by Meisen [149], based on laboratory experiments in which a batch of millimeter size ore particles plus dilute sulfuric acid was spouted with air to achieve rapid extraction of copper from the ore.

13 | *Design*

In this chapter, we shall attempt to summarize how and to what extent the knowledge of spouted bed behavior presented in the preceding chapters can be applied for designing a spouted bed system, in what areas experiments will be necessary for obtaining information relevant to design, what scale-up criteria will be required, and finally what practical considerations will enter the detailed design.

13.1 GENERAL CONSIDERATIONS

The principal information required from design calculations is the size of the spouting vessel and the capacity of the gas blower for a specified process duty. The calculation procedure would depend on the type of process for which the system is to be designed. Let us briefly recapitulate the main design considerations with respect to the various types of processes for which a spouted bed might be used:

(a) In constant rate drying, the drying gas would become saturated, or nearly so, within a vertical distance less than the depth of any practical sized spouted bed, as explained in Chapter 9. The drying capacity will therefore be controlled by the quantity of gas which can be passed through the bed per unit weight of solids, and not by the residence time of the solids in the bed, nor by the nature of the material being dried. The bed size required for a given duty in these circumstances will be determined entirely by spouting parameters, rather than by the kinetics of the drying process.

(b) Similarly, for heating or cooling of solid particles, the approach to equilibrium between gas and solids in a spouted bed is very rapid, as discussed in Chapter 8. Usually, it would be safe to assume here too that in any unit of industrial size, the entire gas leaves the bed at nearly the same temperature as the bulk bed solids; and therefore, design calculations can again be performed without reference to interphase transfer coefficients. For the unusual case, the procedure for verifying the equilibrium assumption has been presented in Chapter 8, indicating how the design will be affected if thermal equilibrium between gas and solids is not attained.

(c) In contrast, for drying in the falling-rate regime, the bed size will be determined by the requirement of solids residence time in the bed, since the drying rate in this case is controlled by the rate of moisture diffusion from within the particles to their surface and not by the moisture-carrying capacity of the drying gas. The design calculations must therefore take into account the nature of the particles being dried (via moisture diffusivity), in addition to the spouting parameters. Detailed information on bed structure and flow patterns is still of no real value, the rigorous cyclical calculations discussed in Chapter 9 being too idealized, in addition to being tedious, from the point of view of engineering design. The calculation method of Becker and Sallans [16], described in Chapter 9, is based on disregarding local variations in drying conditions, and treating the bed simply as a well-stirred mass of isothermal solids. While these assumptions have proved to be well justified in the case of wheat drying under a fairly wide range of conditions of practical interest, their validity cannot be taken for granted when designing for new drying situations where a different type of material, size or geometry of bed, moisture range, drying gas temperature, etc. are involved. Experimental verification of the applicability of this method, and even more so of the more empirical methods in Chapter 9, as outlined in Section 13.2.C, will then be unavoidable for reliable design.

(d) For other solids treatment processes where gas–particle interaction is more complex than in pure drying (granulation, particle coating, shale pyrolysis, etc.) it is difficult, with the present state of knowledge, to recommend any generalized design procedure. The following remarks summarize the difficulties, and might provide some guidance in design:

On a single particle scale, the spouted bed with its nonhomogeneous structure and highly variable flow conditions presents a very complicated picture of gas–particle interaction. Whether or not the simplified model of a well-stirred bed with plug flow of gas, used in the case of diffusion-controlled drying, is adequate as a basis for design would depend on the specific process, and this can be determined only by experiment. If experimental results are correctly described by the model mentioned, the design of a spouted bed system for carrying out that particular process, like the design for diffusion-controlled drying, can be worked out using essentially the same procedure as for a fluidized bed, or indeed as for any well-stirred bed. The usual kinetic models for conversion of solids, such as those presented by Kunii and Levenspiel [110, Chapter 15] for fluidized beds, will then apply to spouted bed design also. If, on the other hand, the process is sensitive to the difference in conditions which prevail between the spout and the annulus—conditions to which a particle is cyclically exposed—then the design problem becomes much more complex. For instance, whereas in drying the brief residence time of a particle in the spout could be safely ignored, the same simplification cannot be made in the case of shale retorting because particle breakdown during this brief period is all-important in determining the overall rate of the retorting process, and must be taken into account in formulating any design equations. Similarly, in both granulation and coating, the distribution of residence time between the spout and the annulus during each cycle, and not only the mean residence time, plays a key role, since here it is essential that the evaporation rate be sufficiently rapid to fully dry the liquid layer deposited on a particle before it gets coated by the next layer. Otherwise, the bed solids would soon become sticky, disrupting the progress of the whole operation.

In the absence of any theoretical models applicable to these more complex processes, the design problem becomes specific to the type of process. Design calculations in such cases would therefore have to rely heavily on kinetic data from bench scale and then pilot plant experiments, with knowledge of bed dynamics providing guidance for scaling up.

(e) Even for the simpler type of process where the general procedure for designing a spouted bed becomes basically the same as for a fluidized bed, design and scale-up calculations for the two systems will differ in one important respect—the fluid flow requirement. The minimum spouting velocity, unlike the minimum fluidization velocity, is dependent on bed size and geometry. According to Eq. (2.38), the volumetric flow rate for spouting, with the ratio D_i/D_c fixed, is directly proportional to $D_c H^{1/2}$, and therefore to the square root of bed volume (for a cylindrical bed). In contrast, the corresponding flow rate in fluidization is independent of bed height and is directly proportional simply to the cross-sectional area of the column.

The above point of difference between spouting and fluidization will manifest itself throughout the design calculation, giving rise in the case of spouting to some unexpected answers when viewed from the vantage point of a fluidized bed designer.

(f) When it comes to processes where conversion of gas rather than of solids is the main objective, the similarity between fluidization and spouting arising from the well-mixed nature of the solids is no longer relevant. The gas flow behavior in the two systems is quite different, and therefore the usual approach, based on bubble dynamics, to the design of fluidized bed reactors obviously does not apply to spouted beds. In this area, the two-region model of a spouted bed presented in Chapter 10 should provide a basis for interpretation of conversion data from laboratory experiments and for subsequent scale-up calculations. It should, however, be stressed that the validity of the above theoretical model still remains to be tested against experimental results.

13.2 EXPERIMENTS REQUIRED AND SCALE-UP CRITERIA

A. Spouting Behavior

Preliminary laboratory tests with the particular material are always necessary to determine spouting behavior. The use of a 15 cm diameter × 1–2 m high transparent column, with a conical base of 60° included angle should be suitable for testing most materials. The arrangement for introducing the gas at the bottom should be such that the size of the inlet orifice can be conveniently varied up to about 5 cm diameter, the limiting D_i/D_c ratio for stable spouting being roughly $\frac{1}{3}$. Some form of bed support, to prevent the solids from draining into the gas approach pipe when the bed is not spouting, will be necessary (see Section 13.3.D).

The information obtainable from preliminary tests, as well as its usefulness, is discussed below.

1. *Spoutability*

Can stable spouting be established in a bed at least 30 cm deep using a 1–2 cm diameter orifice? If spouting is achievable only with a much smaller orifice and a much shallower bed, it is doubtful if stable spouting on a larger scale will at all be possible. For materials with a particle size spread, spoutability will be improved by narrowing the size range (see Chapter 6), if this is permitted by the requirements of the particular process.

2. Particle Attrition

Is particle attrition acceptable? The extent of attrition should be determined, over a period similar to the expected residence time of the solids for the process under consideration, by measuring the amount of dust collected in an overhead cyclone, as well as the change in size distribution of the bed solids. Attrition rate is a strong inverse function of gas inlet diameter and should therefore be determined using the largest orifice size permissible for stable spouting. No scale-up correlations with respect to particle breakdown are available but the factors involved have been qualitatively discussed in Chapter 7. In general, attrition would be less severe on a larger scale, but grossly unacceptable attrition on the laboratory scale could rule out the use of a spouted bed at this stage.

3. Minimum Spouting Velocity

The value of U_{ms} for several bed depths, using at least two orifice sizes, should be determined in order to check the validity of Eq. (2.38) for the material under test. This equation, though found to be valid for a variety of solids (Table 2.4), is nevertheless empirical, and does not include parameters to describe shape and surface characteristics of the solid particles. If the test results do not obey the equation, it might be possible to obtain from these results an empirical correction factor for d_p applicable to the particular material. The corrected value of d_p can then be used in Eq. (2.38) for scale-up calculations. Similarly, an empirical correction for fluid properties might also be necessary if the fluid used is a gas of viscosity and/or density very much different from that of air. A quick solution of Eq. (2.38) may be obtained using a nomograph [188, 262].

The second spouting velocity equation [Eq. (2.42)] discussed in Chapter 2, which is likely to give better predictions for beds larger than 60 cm diameter, also requires experimental evaluation of both the correction factors mentioned above. This equation, unlike Eq. (2.38), does include shape factor and fluid viscosity terms; however, the former is really an experimental correction factor rather than a description of the actual shape of a particle, while the latter term appears in the equation primarily from dimensional considerations and without explicit experimental backing. It should be emphasized that extrapolation of either equation to a column size much larger than 60 cm diameter is likely to underestimate the spouting velocity.

4. *Particle Velocity at Wall*

A vertical profile of the downward particle velocity, as observed against the transparent wall of the test column, can be easily measured using a stopwatch. These measurements, which may be carried out under an arbitrary set of conditions (say $H/D_c = 4$, $D_i/D_c = 1/6$, $U_s = 1.1U_{ms}$), can be interpreted in terms of particle cycle time, solids turnover rate, and solids cross flow rate. Although no reliable scale-up correlations with respect to these properties have been developed, some indication of these aspects of solids movement in larger beds and under different operating conditions can be obtained from the particle velocity at wall observations, with the aid of empirical equations (e.g., Table 3.3) and experimental trends (e.g., Fig. 3.12) given in Chapter 3.

The importance of information on solids movement and the exact use of such data in design calculations will depend on the type of process involved. In designing a continuous drier, for example, all that is required is to ensure that the mean residence time allowed from kinetic considerations is many times higher than the particle cycle time; otherwise, the perfect mixing assumption would not be approximated. In the design of a blender, on the other hand, the batch time (or mean residence time) will itself be determined by the average circulation rate, while in granulation or particle coating, the design must allow sufficient time in the annulus during each cycle for evaporation of moisture from each newly deposited liquid layer.

5. *Maximum Spoutable Bed Depth*

Determination of H_m on the laboratory scale would be useful, again as a means for checking the validity of the predictive correlations presented in Chapter 6, for the specific material under test. The operating bed depth in a column of industrial size (say >60 cm diameter) will in general be well below the maximum spoutable bed depth, which in itself is therefore of little practical interest. The ratio H/H_m is, however, a useful scale-up parameter, since certain aspects of flow behavior in beds of different sizes operating at the same ratio of H/H_m have been found to be similar, as recorded in Chapters 2, 3, and 5.

B. Rate and Equilibrium Data

The same type of information on transfer or reaction rates and on the relevant driving forces is required here as for designing any fluid–solid system. Methods for measuring the rate of reaction, determining the rate-controlling step, and interpreting laboratory results in terms of a realistic kinetic model, for various types of noncatalytic fluid–solid reactions as well as for catalytic reactions, are discussed in Levenspiel's textbook [118], and with particular reference to fluidized beds in the book by Kunii and Levenspiel [110].

Briefly, for a continuously operating spouted bed type of system, the full progression of experimental work would be thin-layer tests under a simulated spouted bed environment; batch tests in a spouted bed; continuous laboratory scale experiments; and finally continuous runs on the pilot plant scale. Some of these stages might be waived in practice, depending upon how seriously the overall process is affected by local variations within the bed and how close a design is warranted by economic factors. It should, however, be recognized that the various kinetic models discussed in the books cited above, despite their sophistication, are based on fairly idealized behavior, and can be expected to apply only to relatively simple situations. Levenspiel [118, p. 357] has noted that for a fluid–solid reactor ". . . where the kinetics are complex and not well known, where the products of reaction form a blanketing fluid phase, where temperatures within the system vary greatly from position to position, analysis of the situation becomes difficult and present design is based largely on the experiences gained by many years of operations . . . ." A spouted bed for carrying out processes involving similar complexities will be no exception.

C. Pilot Plant Work

With the spouting behavior determined and preliminary rate and equilibrium information obtained, it should be possible to work out an *approximate* design of the full-scale spouting unit, estimating the hydrodynamic data required (e.g., spouting velocity, gas and solids flow patterns, spout diameter, spout voidage) from information given in Chapters 2–5. The calculation procedure for operations involving treatment of solids by a gas—physical operations as well as noncatalytic chemical processes (as classified in Chapter 11)—will be similar to that outlined in Chapters 8 and 9, while for vapor phase chemical reaction, calculations along the lines indicated in Chapter 10 will be required. Presuming that on the basis of this approximate design, the spouted bed continues to stay in the running on both technical and economic grounds vis-à-vis other systems which might be under consideration, the next phase of experimental work would usually be carried out on a 30–60 cm diameter pilot unit.

1. *Process Rate*

The main purpose of pilot plant work will be to check the validity, over a range of operating conditions, of any previously determined kinetic and equilibrium data or, in the case of complex processes (e.g., those involving particle growth, shrinkage, breakdown, chemical change, elutriation, recycle) to determine these data afresh. A few spot checks is all that might be required to achieve the former objective, while a more extensive program of experiments will be necessary in the latter case. Design calculations for the full-scale unit, based on optimum process conditions as established on the pilot plant, can then be performed with greater confidence, using the same procedures as employed previously for the approximate design.

2. *Operational Stability*

The pilot plant also provides the opportunity of establishing operational stability under process conditions, and a few continuous runs of long duration are recommended for this purpose. In particular, any evidence of agglomeration or caking should be carefully watched, since the formation of even a few large persistent lumps in the bed, or of a hard cake on the wall, especially in the vicinity of the gas inlet, can have a disastrous effect on spouting stability. In experiments on drying of ammonium nitrate [96], it was found that these difficulties tended to snowball, an interruption in the smooth flow of solids causing a rise in temperature in the lower part of the bed, which aggravated agglomeration and caking. The problem in this case appeared to have been caused by a rather critical interaction between moisture content and temperature of the bed solids, and was overcome by manipulation of operating conditions.

3. *Product Quality*

The question of product quality might also require detailed attention at the pilot plant stage. In a drying process, for instance, the operating bed temperature must obviously remain well below the fusion, decomposition, or charring temperature of the solids, but even a lower limit on bulk solids temperature might be placed by considerations of product quality. Such was found to be the case in spouted bed drying of wheat [16] and of ammonium nitrate [96], the bread-making quality of the former and the explosive properties of the latter being both apparently sensitive to the temperature–moisture history of the particles, as well as to their rate of drying. Determination of the upper safe limit of solids temperature in these cases was important, because for maximum capacity it was desirable to operate the spouted bed at the highest permissible bed temperature. In processes of this type, monitoring the quality of the pilot plant product, including field testing if necessary, might have to be carried out as a routine in order to establish the limits of safe operating conditions. Whether or not the same limits will apply in larger-scale operation will depend on the detailed mechanism by which the specific damage occurs, but at least pilot plant results will provide the minimum reassurance required on this count for proceeding to the next larger scale.

13.3 SOME PRACTICAL SUGGESTIONS

The following additional features of design, though secondary to sizing of the main spouting column and the gas blower, are nevertheless important for successful operation of a spouted bed system. The best we can do here is to offer some practical suggestions, based partly on information from published literature and partly on experience.*

A. Gas Entrance

It is common practice to introduce the spouting gas into the base of the bed through a calming section in order to ensure that any flow instabilities are damped out before the gas enters the bed. A straight vertical approach section of about 10–12 pipe diameters, with a bundle of thin-walled tubes inside the pipe to serve as straightening veins, has often been used, as in the case of an orifice flowmeter [177, §5, p. 11], although how critical this requirement is for a spouted bed has never been established. The approach pipe itself may be either of the same diameter as the required gas orifice size, or it may be larger, narrowing down to the desired size at the point of its entry into the bed. The reduction in size may alternately be effected by insertion of an orifice plate, as in Fig. 6.2a, with the advantage of convenience in changing the aperture size—an important consideration in an experimental unit. The high pressure loss associated with an abrupt contraction is, however, a disadvantage; hence, for a permanent installation the use of a simple converging approach pipe (Fig. 13.1a) or of a converging nozzle (Fig. 13.1b) is more suitable. Up to a point, graduating the con-

---

*Our own, as well as that of W. S. Peterson of the National Research Council of Canada, to whom we are grateful for his contribution to this section and for his comments on this chapter.

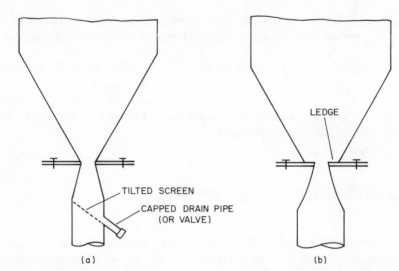

Fig. 13.1. Gas entrance arrangements: (a) Converging approach pipe; (b) converging nozzle approach.

vergence will lower the pressure loss, but this constitutes a practical advantage only within limits, since for too gradual a convergence the pipe length will become inordinately large. Pressure loss with a nozzle-shaped convergence is smaller than with straight convergence. The measured loss across a 1.6 cm diameter gas inlet at an air flow rate of 1.8 m³/sec, with the entrance of Fig. 13.1b, was about 50% lower than with a 65° converging approach pipe of the type depicted in Fig. 13.1a [180]. Other aspects of spouting behavior, namely, minimum spouting velocity and particle velocity at wall, were found to be only marginally affected by the use of a converging entrance, regardless of whether the convergence occurred gradually or abruptly.

In addition to the foregoing considerations, the design of the gas entrance must also take into account the effect on spouting stability, discussed in Chapter 6.

B. Feed and Discharge of Solids

In continuous operation, the solids are commonly fed into the top of the bed near the column wall and are discharged through an overflow pipe on the opposite side, which serves to maintain the bed level. A feed pipe reaching down to the bed surface or just below it may be used to direct the flow of solids from the discharge end of the feeder; alternatively, the solids may be allowed to rain down freely from the feeder near the column wall.

The former arrangement (see Fig. 1.3a) avoids short circuiting of fresh feed through the overflow and is therefore necessary in a narrow column. For a wide column, the latter arrangement is more suitable since it eliminates the possibility of choking the feed pipe, which can be particularly troublesome if the solids are not free flowing. Another problem which may arise is the blowback of gas through the feeder, which, apart from causing a dust nuisance, can also lead to the more serious difficulty of solids caking in the feeder, as it did in the case of ammonium nitrate drying [96]. In such cases, the use of a star type feeder will minimize blowback. A more effective solution of this particular difficulty, as well as of the more general problem of dust nuisance, is to put the entire spouting column under a slight suction by discharging the exit gas through some suction device, e.g., exhaust fan, water ejector, or steam ejector. Provision of a cyclone separator on the upstream side of the suction device will normally be necessary. If an exhaust fan is used, an additional gas-cleaning device such as a bag filter or scrubber might be needed at the final discharge point.

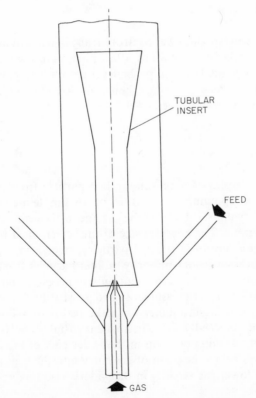

Fig. 13.2. Solids feed scheme developed by Baticz *et al.* [13].

Solids feed can also be introduced at the bottom rather than at the top, by two different methods. The first method, depicted in Fig. 1.3b, simply involves entrainment of the feed particles into the spouting gas before the gas enters the bed. Spouting stability is reported to have been improved by such a feeding arrangement, but particle attrition was greater than with top feeding [134]. The second method, developed by Baticz *et al.* [13] with particular concern for avoiding wear and tear on the particles, is shown schematically in Fig. 13.2. The main objective of their feeding technique was apparently the elevation of friable coal particles in the spouted bed part of their Ecosorber unit (shown in Fig. 11.10). Advantage was taken of the suction effect created by the gas jet entering the bed, but in order to achieve the high feed rates required, it was necessary to accentuate the suction by using a tubular insert surrounding the spout region, as illustrated in Fig. 13.2. Feed rates achieved with the insert were 8–10 times those without the insert, at corresponding gas flow rates.

C. Bed Level Control

The need for any special level control arrangement will arise only when discharging the solids through an overflow pipe, as in Fig. 1.3, is not desirable. In such cases, the hot-wire-probe level detector developed by Osberg [170] for fluidized beds, or some similar device, should be suitable for automatic control of spouted bed level.

D. Bed Support

Provision is necessary for preventing bed particle from falling into the gas pipe when the spouting gas is turned off. A simple method, commonly used in laboratory units, is to cover the gas inlet orifice with a coarse screen, as shown in Fig. 6.2. The aperture size of the screen must be only slightly smaller than the majority of the bed particles, since bridging of the solids above the screen restrains the minority of finer particles from draining into the pipe. The use of a fine screen is not only unnecessary but also undesirable, since it gives a higher pressure loss and gets easily clogged by any fines present, thereby causing instability when the bed is subsequently spouted. A better scheme, especially for a permanent installation, is to locate the coarse screen in a sloping position in the wider part of the approach pipe a few centimeters below the point of gas entry into the bed, as indicated in Fig. 13.1a. The lower gas velocity in the wider section makes for economy

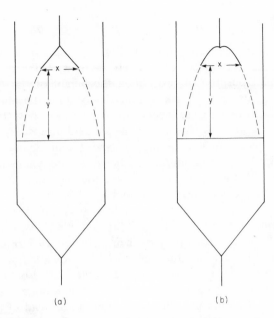

Fig. 13.3. Spout deflectors used by Peterson [180]: $x \simeq \frac{1}{2}D_c$; $y \simeq D_c$. (a) Simple conical deflector; (b) shaped deflector.

in pressure loss, while the sloping position of the screen allows particles entrapped in the approach pipe to slide into the capped side opening shown in the figure, from where they can be periodically removed. This opening also provides a convenient access for cleaning the screen if it does get blinded. Such an arrangement has proved to be very successful in commercial grain driers [180].

The insertion of a conical plug from above into the gas inlet orifice has been proposed as an alternate method for supporting the bed solids [179]. Since support is required only when the bed is not spouting, the conical plug can be gradually withdrawn during start-up as the gas flow is increased, while at the time of shutting down, the plug must be reinserted before the gas supply is turned off. This technique should be particularly suitable for processes where caking of the solids on the supporting screen poses a problem, although some form of built-in safeguard against unexpected failure of gas supply during operation will be necessary. Other suggested alternatives to a supporting screen include a special poppet valve in the inlet pipe which will not open until the pressure reaches a predetermined value, and a quick-acting valve which will bring the gas flow rate from zero up to that required for spouting over a very short time interval [34].

E. Spout Deflector

Spout deflectors of the type shown in Fig. 13.3 (also visible in the photograph of Fig. 12.1) have been effectively used by Peterson [179, 180] to achieve the dual purpose of restraining the height of the spout and of inducing the solids to fall back symmetrically onto the annular region. The former function enables operation at high gas flow rates without entrainment and prevents blowout of the solids during start-up, while the latter function helps in positioning the spout centrally and in keeping spout wandering to a minimum. A smoother and more symmetrical fountain was obtained with the shaped deflector of the type shown in Fig. 13.3b than with the simple conical deflector of Fig. 13.3a.

A spout deflector can also serve as a solids breaker if the bed is operated at a high spouting velocity. Such a "breaker plate" was employed by Klimenko and Rabinovich [101] in their manganese chloride drier, and should be useful for accentuating the normal solids attrition in processes where attrition is a beneficial feature (see Table 12.1). For other types of processes requiring high operating gas flow rates but not excessive particle breakdown, a restraining net above the fountain, instead of a solid plate deflector or breaker, has been used to minimize carryover [201, p. 236].

Appendix

NOMENCLATURE

A	Area
A_a	Cross-sectional area of annulus
A_c	Cross-sectional area of column
A_e	Cross-sectional area of piezoelectric sensing element
A_i	Cross-sectional area of fluid inlet orifice
A_p	Surface area of particles
A_p'	Surface area of a single particle
A_s	Cross-sectional area of spout
A_w	Area of heat transfer surface [Eq. (8.12)]
B	Empirical constant in Eq. (3.20)
C	Concentration of diffusing vapor or reactant in gas
C^*	Equilibrium value of C
C_a	Concentration of unconverted reactant in annulus gas
C_{aH}	C_a at $z = H$
C_b	Fractional concentration of traced solids in bed
C_{b0}	C_b at $t = 0$
C_D	Drag coefficient, drag force$/[(\frac{1}{4}\pi\, d_p{}^2)(\frac{1}{2}\rho_f U^2)]$
C_e	Concentration of tracer or reactant in total gas leaving bed
C_f	Fractional concentration of traced solids in feed
C_i	Concentration of tracer, reactant, or vapors in gas entering bed

C_0	Concentration of tracer in gas within bed at $t \leq 0$ (Fig. 3.5)
C_p	Fractional concentration of traced solids in product
C_s	Concentration of reactant in spout gas
$C_{sH'}$	C_s at $z = H'$
C_v	Volatile matter content of carbonizing particles
C_1, C_2	Constants in Eq. (3.12)
ΔC_{Lm}	Logarithmic mean concentration driving force
c_i, c_i', c_i''	Inertial force constants in Eqs. (2.3), (2.4), and (2.5)
c_{Pg}	Heat capacity of gas
c_{Pp}	Heat capacity of a particle
c_{Ps}	Heat capacity of solids
c_v, c_v', c_v''	Viscous force constants in Eqs. (2.3), (2.4), and (2.5)
$\mathcal{D}$	Internal moisture diffusivity of a particle
D	Superficial axial dispersion coefficient for the annulus in terms of the diffusion model
D_b	Diameter of upper surface of bed
D_c	Column diameter
D_i	Fluid inlet diameter
D_s	Spout diameter
D_v	Diffusivity of vapor in gas
d_p	Particle diameter
d_{pi}	Particle diameter of size fraction x_i
d_v	Diameter of sphere of same volume as particle
$E(\theta)$	Exit age distribution function
e	Coefficient of restitution
F	Volumetric feed rate
$F(\theta)$	Dimensionless tracer response to a step input = output tracer concentration/ input tracer concentration
F_r, F_z	Forces acting on an element of particles in the spout wall (Fig. 5.2a)
f	Friction factor
	Function
G	Fluid mass flow rate per unit of column cross section
G_{ms}	G at minimum spouting
g	Acceleration of gravity
H	Bed depth
H'	Value of z at top of fountain
H_f	Fluidized bed height
H_m	Maximum spoutable bed depth
h_p	Heat transfer coefficient between fluid and particle
$\bar{h}_p$	Composite fluid-to-particle heat transfer coefficient [Eq. (8.11)]
h_s	Heat transfer coefficient between submerged object and bed
h_w	Heat transfer coefficient between wall and bed, surface-mean value
$I(\theta)$	Internal age distribution function
K, K', k	Proportionality constants
K_p	Mass transfer coefficient between particles and fluid
$\bar{K}_p$	Composite particle-to-gas mass transfer coefficient
K_r	Reaction rate constant based on volume of solids [Eq. (10.4)]
K_r'	Reaction rate constant based on external surface area of solids (Eq. (10.4a)]
k_b	Effective thermal conductivity of bed
k_g	Thermal conductivity of gas

k_p	Thermal conductivity of solid particle
L	Latent heat of vaporization plus heat of desorption
l	Characteristic length of confinement [Eq. (2.3)]
M	Momentum gained by particles or lost by fluid jet [Eq. (2.39)]
m	Mass of element of particles in the spout wall, in Chapter 5
	Local moisture content within a particle, in Chapter 9
$\bar{m}$	Final moisture content of solids, dry basis
$\bar{\bar{m}}$	Average moisture content of a statistical population of drying particles, dry basis
m_e	Moisture content of exit solids, dry basis
m_e'	Moisture content of exit solids, wet basis
m_i	Moisture content of inlet solids, dry basis
m_0	Initial moisture content of solids, dry basis
m_p	Mass of a particle
m_s	Mass flow rate of particles in the spout at any level, in Chapter 5
	Surface moisture content of particles, dry basis, in Chapter 9
N	Number of collisions per unit time
	Number of moles of reactant
	Number of stages
n	Number of particles accelerated per unit time, in Chapter 2
	State-of-flow index
n_z	Cumulative number of rising particles in the spout at bed level z
P	Pressure at any point
P_b	Downward force per unit cross-sectional area of the annulus
P_0	Static pressure at no flow
$-\Delta P$	Pressure drop
$-\Delta P_F$	Fluidized bed pressure drop
$-\Delta P_M$	Maximum pressure drop prior to onset of spouting (Fig. 2.1)
$-\Delta P_s$	Spouted bed pressure drop
$(\Delta P_s)_{max}$	ΔP_s for $H = H_m$
Q_H	Volumetric flow rate required for spouting a bed of depth H
Q_s	Volumetric flow rate of fluid in spout
q	Heat transfer rate
R	Recycle ratio (Fig. 4.13)
	Gas constant [Eq. (9.7)]
$R_i(R_1, R_2, \ldots)$	Recycle ratio for CST i (Fig. 4.15a)
r	Radial distance from axis
	Reaction rate, rate of disappearance of reactant [Eqs. (10.4), (10.4a)]
r_c	Column radius
r_p	Particle radius
r_s	Spout radius
$S_i(S_0, S_1, S_2, \ldots)$	Weight of solids in CST i (Figs. 4.12 and 4.15a)
s	Coefficient defined by Eq. (2.43)
T	Temperature
T_{as}	Adiabatic saturation temperature
T_b	Bulk bed solids temperature
T_{b0}	T_b at $t = 0$
T_g	Gas temperature
T_{ge}	Exit gas temperature
T_{gi}	Inlet gas temperature

T_{gw}	Wet bulb temperature
T_0	Uniform bed temperature outside thermal boundary layer at wall
T_p	Particle temperature
T_{p0}	T_p at $t = 0$
T_{si}	Inlet solids temperature
T_{se}	Exit solids temperature
T_w	Wall temperature
ΔT_{Lm}	Logarithmic mean temperature driving force
t	Time
$\bar{t}$	Mean residence time of solids, W_b/W_s
t_c	Maximum particle cycle time in bed
$\bar{t}_i$	Mean residence time of solids in one stage of a multistage system
U	Superficial fluid velocity
U_a	Upward superficial fluid velocity in the annulus
U_{aH}	U_a at $z = H$
U_f	Superficial fluidization velocity
U_M	Superficial fluid velocity corresponding to ΔP_M (Fig. 2.1)
U_m	U_{ms} at $H = H_m$
U_{mf}	Minimum superficial velocity for fluidization
U_{ms}	Minimum superficial fluid velocity for spouting
U_r	Volumetric rate of radial fluid percolation per unit area of spout–annulus interface
U_s	Superficial spouting fluid velocity
$\bar{U}_s$	Volumetric upward fluid flow rate through spout per unit of spout cross-sectional area
$\bar{U}_{sH'}$	$\bar{U}_s$ at $z = H'$
u	Fluid velocity
u_i	Fluid velocity through inlet orifice
$(u_i)_{ms}$	u_i at minimum spouting condition
u_s	Upward interstitial fluid velocity in the spout
u_{sH}	u_s at $z = H$
V	Bed volume
V_p	Plug flow volume
$V_p{}'$	Volume of a single particle
V_r	Volume of solids in the reaction zone
v	Upward particle velocity in spout
v_a	Upward velocity of spout particle adjacent to spout wall
v_m	Radial mean upward particle velocity in spout
v_0	v at $r = 0$
v_r	Radial velocity of particle leaving annulus
v_t	Free fall terminal velocity of a particle
v_w, v_z	Downward particle velocity at column wall
W	Mass downflow rate of solids in annulus = mass upflow rate of solids in spout = mass circulation rate of solids past any horizontal plane in bed
W_b	Weight of bed
$W_i(W_1, W_2, \ldots)$	Mass flow rate of solids out of CST i (Fig. 4.15a)
W_{max}	W at $z = H$ = maximum mass circulation rate of solids
W_s	Mass flow rate of solids into or out of bed

$\Delta W/\Delta z$	Mass rate of solids cross flow per unit of bed height in the cylindrical portion of the bed
X	Fractional overall conversion, $1 - C_e/C_i$
x_e	Equilibrium tracer concentration in bed, weight fraction
x_i	Mass fraction of particles of size d_{pi}
x_0	Tracer concentration at top of bed, weight fraction
x_n	Tracer concentration at bottom of bed
Y	Lateral distance from wall (Fig. 8.3)
z	Vertical distance from fluid inlet
Z	Vertical distance from top of heated zone (Fig. 8.3)
α	Thermal diffusivity of particle, $k_p/\rho_s c_{Pp}$
α_b	Thermal diffusivity of bed, $k_b/\rho_b c_{Ps}$
β	Gas–solids interaction factor in Eqs. (4.14) and (4.15)
γ	Angle of internal friction
ϵ	Voidage (1 − volumetric fraction of solids)
ϵ_a	Voidage in the spouted bed annulus
ϵ_{mf}	Voidage at minimum fluidization
ϵ_0	Loose packed bed voidage
ϵ_s	Spout voidage
θ	Included angle of cone
	Dimensionless time
λ	Particle shape factor in Eq. (6.5), surface are of particle/surface area of equi-volume sphere = $0.205 A_p'/(V_p')^{2/3}$
μ	Fluid viscosity
ρ_b	Solids bulk density
ρ_{bs}	Solids bulk density in the spout
ρ_f	Fluid density
ρ_g	Gas density
ρ_s	Particle density
τ	Interparticle shear stress
τ_s	Upward shear stress at spout wall
τ_w	Upward shear stress on annular solids at vessel wall
ϕ	Angle of repose
Ω	Net downward force per unit volume exerted by the annular solids
ψ	Arbitrary particle shape factor in Eq. (2.35)
$\overline{\psi}$	Particle shape factor in Eq. (2.26), reciprocal mean diameter by screen analysis, d_p/diameter of equivolume sphere, d_v
ψ'	Particle shape factor in Eq. (9.17), $6V_p'/A_p' d_v$

Dimensionless Groups

Ar	Archimedes number, $g d_p^3 \rho_f(\rho_s - \rho_f)/\mu^2$
Bi_H	Heat transfer Biot number, $h_p r_p/k_p$
Bi_M	Mass transfer Biot number, $K_p r_p/\mathscr{D}$
Eu	Euler number, $-\Delta P/\rho_f u^2$
Fo_H	Heat transfer Fourier number, $\alpha t/r_p^2$
Fo_M	Mass transfer Fourier number, $\mathscr{D}t/r_p^2$
Fr	Froude number, $U^2/g d_p$

Gu	Gukhman number, $(T_{ge} - T_{gw})/T_{ge}$
j_H	Heat transfer factor, $St \cdot Pr^{2/3}$
j_M	Mass transfer factor, $St_M \cdot Sc^{2/3}$
Ko	Kossovich number, $Lm_i/c_{ps}(T_{gi} - T_{si})$
$\bar{M}$	$(\bar{m} - m_s)/(m_0 - m_s)$
$\bar{\bar{M}}$	$(\bar{\bar{m}} - m_s)/(m_0 - m_s)$
Nu	Nusselt number, $h_p d_p/k_f$
Pr	Prandtl number, $c_{pf}\mu/k_f$
Re	Superficial particle Reynolds number, $d_p U \rho_f/\mu$
Re_i	Orifice Reynolds number, $d_p u_i \rho_f/\mu$
Re_m	Re at $U = U_m$
Re_M	Re at $U = U_M$
$(Re_i)_{ms}$	Re_i at $u_i = (u_i)_{ms}$
Sc	Schmidt number, $\mu/\rho_f D_v$
Sh	Sherwood number, $K_p d_p/D_v$
St	Heat transfer Stanton number, $\bar{h}_p/c_{pf} U_s \rho_f$
St_M	Mass transfer Stanton number, $\bar{K}_p/U_s$
$\mathcal{X}$	$St \cdot A_p/A_c$
$\mathcal{X}'$	$St_M \cdot A_p/A_c$
x	Fractional temperature approach to equilibrium [Eq. (8.11b)]
x'	Fractional concentration approach to equilibrium [Eq. (9.14)]
χ	$(A_p'/V_p')(\mathcal{D}t)^{1/2}$
$\bar{\chi}$	$(A_p'/V_p')(\mathcal{D}\bar{t})^{1/2}$

Abbreviations

BASF	Badische Analin und Soda Fabrik	NRC	National Research Council of Canada
CST	Completely stirred tank	PEC	Potasse et Engrais Chimiques
H-iron	Iron ore reduction process of Hydrocarbon Research Inc.	RTD	Residence time distribution
ICI	Imperial Chemical Industries	TCC	Thermofor catalytic cracking process
IEL	Indian Explosives Limited	UBC	University of British Columbia

CONVERSION FACTORS

1 m	3.281 ft		1 N-sec/m²	0.672 lb/ft-sec
1 m²	10.76 ft²		1 J	0.738 ft-lb$_f$
1 m³	35.32 ft³		1 kJ	0.948 Btu
1 kg	2.205 lb		1 MJ	0.372 hp-hr
1 Mg	0.984 long ton		1 kJ/m³	0.0268 Btu/ft³
1 Mg/m³	62.43 lb/ft³		1 kJ/kg	0.430 Btu/lb
1 kN/m²	0.145 lb$_f$/in.²		1 kJ/kg-°K	0.239 Btu/lb-°F
1 gm/sec	7.936 lb/hr		1 W	3.412 Btu/hr
1 kg/sec	3.543 long ton/hr		1 W/m²	0.317 Btu/hr-ft²
1 m³/sec	2119 ft³/min		1 W/m-°K	0.578 Btu/hr-ft-°F
1 gm/sec-m²	0.737 lb/hr-ft²		1 W/m²-°K	0.176 Btu/hr-ft²-°F

References

1. Abdelrazek, I. D., An analysis of thermo-chemical deposition in spouted beds. Ph.D. thesis, Univ. of Tennessee, Knoxville, 1969.
2. Ageyev, F. G., Soroko, V. E., and Mukhlenov, I. P., Contact apparatus with a jet circulative bed of granular material. *Khim. Prom.* (*Moscow*) **46**, 465 (1970).
3. Aggor, L., and Fritz, W., Kinetics of pyrolytic deposition of silicon carbide on spherical particles in a fluid bed reactor. *Chem. Ing. Tech.* **43**, 472 (1971).
4. Ashton, M. D., and Valentin, F. H. H., The mixing of powders and particles in industrial mixers. *Trans. Inst. Chem. Eng.* **44**, T166 (1966).
5. Auf, A. A., Romankov, P. G., and Frolov, V. F., Investigation of drying of certain granular materials in a spouted bed. *Zh. Prikl. Khim.* (*Leningrad*) **39**, 1724 (1966).
6. Baerns, M., Effect of interparticle adhesive forces on fluidization of fine particles. *Ind. Eng. Chem. Fundam.* **5**, 508 (1966).
7. Bardin, I. P., Vavilov, N. S., Gess-de-Kal've, B. A., Diev, V. E., Emel'yanov, V. I., Kanavets, P. I., Melent'ev, P. N., Rumakina, M. A., and Tsylev, L. M., Reduction of iron from ore–fuel pellets in a spouting fluidized bed. *Russ. Met. Fuels* **No. 5**, 13 (1960).
8. Barton, R. K., Rigby, G. R., and Ratcliffe, J. S., Fluids–solids contacting in spouted beds—A review. *Mech. Chem. Eng. Trans.* **4**, 95 (1968).
9. Barton, R. K., Rigby, G. R., and Ratcliffe, J. S., The use of a spouted bed for the low temperature carbonization of coal. *Mech. Chem. Eng. Trans.* **4**, 105 (1968).
10. Barton, R. K., and Ratcliffe, J. S., The rates of devolatilization of coal under spouted bed conditions. *Mech. Chem. Eng. Trans.* **5**, 35 (1969).
11. Baskakov, A. P., Antifeev, V. A., and Lummi, A. P., Study of local heat exchange in a spouted bed using a thermal probe. *Inzh. Fiz. Zh.* **10**, 16 (1966).

12. Baskakov, A. P., and Pomortseva, A. A., A study of hydrodynamics and heat exchange of a spouting bed in conical sets. *Khim. Prom.* (*Moscow*) **46**, 860 (1970).

13. Baticz, A., Pallai, E., and Nemeth, J., Investigation of injector effect and particle movement in an apparatus with spouting bed. Res. Inst. for Tech. Chem., Hung. Acad. of Sci., Budapest, 1969.

14. Beatty, R. L., Carlsen, Jr., F. A., and Cook, J. L., Pyrolytic-carbon coatings on ceramic fuel particles. *Nucl. Appl.* **1**, 560 (1965).

15. Becker, H. A., An investigation of laws governing the spouting of coarse particles. *Chem. Eng. Sci.* **13**, 245 (1961).

16. Becker, H. A., and Sallans, H. R., On the continuous, moisture diffusion-controlled drying of solid particles in a well-mixed, isothermal bed. *Chem. Eng. Sci.* **13**, 97 (1961).

17. Becker, H. A., and Isaacson, R. A., Wheat drying in well-stirred-batch and continuous-moving-bed dryers. *Can. J. Chem. Eng.* **48**, 560 (1970).

18. Beek, W. J., Mass transfer in fluidized beds. *In* "Fluidization" (J. F. Davidson and D. Harrison, eds.), Chapter 9. Academic Press, New York, 1971.

19. Berquin, Y. F., A new granulation process and its application in the field of fertilizer manufacture. *Génie Chim.* **86**, 45 (1961).

20. Berquin, Y. F., Method and apparatus for granulating melted solid and hardenable fluid products. U.S. Patent No. 3,231,413 to PEC, Paris, 1966 (filed 1962). Equivalent Brit. Patent No. 962,265; 1964.

21. Berquin, Y. F., Concentration of phosphoric acid using the "Perlomatic". Rep. No. LTE/70/4. Received by courtesy of Mr. Berquin, PEC, Paris.

22. Berti, L., Pyrolysis of oil shale in a spouted bed. M.S. thesis, Colorado School of Mines, Golden, Colorado, 1966.

23. Berti, L., Operational criterion of a spouted-bed oil shale retort. Ph.D. thesis, Colorado School of Mines, Golden, Colorado, 1968.

24. Beutler, H., and Beatty, R. L., Pyrolytic-carbon coating studies. USAEC Rep., ORNL-3885. Oak Ridge Nat. Lab. Oak Ridge, Tennessee, 1965.

25. Bokros, J. C., The structure of pyrolytic carbon deposited in a fluidized bed. *Carbon* (*Oxford*) **3**, 17 (1965).

26. Botterill, J. S. M., Butt, M. H. D., Cain, G. L., and Redish, K. A., The effect of gas and solids thermal properties on the rate of heat transfer to gas-fluidized beds. *Proc. Eindhoven Fluidzn. Symp., 1967,* p. 442. Netherlands Univ. Press, Amsterdam.

27. Bowers, R. H., Stevens, J. W., and Suckling, R. D., Mixing apparatus. Brit. Patent No. 855,809 to ICI Ltd., London, 1960 (filed 1959).

28. Bowling, K. McG., and Watts, A., Determination of particle residence time in a fluidized bed: Theoretical and practical aspects. *Aust. J. Appl. Sci.* **12**, 413 (1961).

29. Bowling, K. McG., and Waters, P. L., Carbonization of coal in a fluidized bed. *Brit. Chem. Eng.* **7**, 98 (1962).

∨30. Bridgwater, J., and Mathur, K. B., Prediction of spout diameter in a spouted bed—A theoretical model. *Powder Technol.* **6**, 183 (1972).

31. Brinn, M. S., Friedman, S. J., Gluckert, F. A., and Pigford, R. L., Heat transfer to granular materials. *Ind. Eng. Chem.* **40**, 1050 (1948).

32. Brown, R. L., and Richards, J. C., "Principles of Powder Mechanics." Pergamon, Oxford, 1970.

33. Buchanan, R. H., and Manurung, F., Spouted bed low-temperature carbonization of coal. *Brit. Chem. Eng.* **6**, 402 (1961).

34. Buchanan, R. H., and Wilson, B., The fluid-lift solids recirculator. *Mech. Chem. Eng. Trans.* **1**, 117 (1965).

35. Caldas, I., Ph.D. thesis, Univ. of Cincinnati, Cincinnati, Ohio, 1955. Quoted by Ghosh and Osberg [76].

36. Charlton, B. G., Morris, J. B., and Williams, G. H., An experimental study of spouting beds of spheres. Rep. No. AERE-R4852. U.K. At. Energy Authority, Harwell, 1965.
37. Chatterjee, A., Spout-fluid bed technique. *Ind. Eng. Chem. Process Des. Develop.* **9**, 340 (1970).
38. Chatterjee, A., Effect of particle diameter and apparent particle density on internal solid circulation rate in air-spouted beds. *Ind. Eng. Chem. Process Des. Develop.* **9**, 531 (1970).
39. Cholette, A., and Cloutier, L., Mixing efficiency determinations for continuous flow systems, *Can. J. Chem. Eng.* **37**, 105 (1959).
40. Clary, B. L., Agrawal, K. K., and Nelson, G. L., Simultaneous heat and mass transfer from peanuts in a spouted bed. *Meeting Amer. Soc. of Agr. Eng., Chicago, 1970*, Paper No. 70-308. ASAE, St. Joseph, Michigan.
41. Cominco Ltd., Trail, B. C., Canada. Personal communication from B. McDonnell, 1971; Anderson, R. W., Nitrogen-phosphorous fertilizer processes. Rep. dated 1966.
42. Cowan, C. B., Peterson, W. S., and Osberg, G. L., Drying of wood chips in a spouted bed. *Pulp Paper Mag. Can.* **58**, No. 12, 139 (1957).
43. Cowan, C. B., Peterson, W. S., and Osberg, G. L., Spouting of large particles. *Eng. J.* **41**, 60 (1958).
44. Crank, J., "The Mathematics of Diffusion," Chapter 6. Oxford Univ. Press (Clarendon), London and New York, 1956.
45. Danckwerts, P. V., Continuous flow systems: Distribution of residence times. *Chem. Eng. Sci.* **2**, 1 (1953).
46. Davidson, J. F., and Harrison, D., "Fluidized Particles." Cambridge Univ. Press, London and New York, 1963.
47. Davidson, J. F., Robson, M. W. L., and Roesler, F. C., Drying of granular solids subjected to alternating boundary conditions. *Chem. Eng. Sci.* **24**, 815 (1969).
48. Drew, T. B., Mathematical attacks on forced convection problems: A review. *Trans. A.I.Ch.E.* **26**, (1931).
49. Dumitrescu, C., and Ionescu, D., The spouted bed, an aspect of the fluidized bed. *Rev. Chim. (Bucharest)* **18**, 552 (1967).
50. Dumitrescu, C., and Ionescu, D., Calculation of minimum spouting velocity. *Rev. Chim. (Bucharest)* **20**, 697 (1969).
51. Dumitrescu, C., and Ionescu, D., Scale up of a spouted bed installation on the basis of similitude principles. *Rev. Chim.* **21**, 491 (1970).
52. Dumitrescu, C., and Ionescu, D., Contributions to the spouted-bed studies as an aspect of fluidization. *Int. Congr. Chem. Eng. (CHISA), 4th, Prague, September 1972*, Paper No. E5.10. Czechoslovak Society for Industrial Chemistry.
53. Du Pont de Nemours and Co., Wilmington, Delaware, U.S.A., Personal communication, 1968.
54. Eastwood, J., Matzen, E. J. P., Young, M. J., and Epstein, N., Random loose porosity of packed beds. *Brit. Chem. Eng.* **14**, 1542 (1969).
55. Elperin, I. T., Zabrodsky, S. S., Yefremtsev, V. S., and Mikhailik, V. D., On the best construction of sets for spouting beds. Collected papers on "Intensification of Transfer of Heat and Mass in Drying and Thermal Processes," p. 323. Nauka i Tekhnika BSSR, Minsk, 1967.
56. Elperin, I. T., Yefremtsev, V. S., and Dolidovich, A. F., The effect of velocity pulsations on the interfacial heat transfer and structure of a spouting bed. *Heat Transfer Sov. Res.* **1**, 23 (1969).
57. Epstein, N., Correction factor for axial mixing in packed beds. *Can. J. Chem. Eng.* **36**, 210 (1958).
58. Epstein, N., Void fraction variation in the spouted bed annulus. *Ind. Eng. Chem. Process Des. Develop.* **7**, 158 (1968).

59. Epstein, N., and Mathur, K. B., *Chem. Eng. (London)* **No. 232,** CE 382 (1969).
60. Epstein, N., and Mathur, K. B., Heat and mass transfer in spouted beds—A review. *Can. J. Chem. Eng.* **49,** 467 (1971).
61. Ergun, S., Fluid flow through packed columns. *Chem. Eng. Progr.* **48,** No. 2, 89 (1952).
61a. Fakhimi, S., and Harrison, D., Multi-orifice distributors in fluidized beds: A guide to design. *"Chemeca 70" Chem. Eng. Conf., Australia, 1970,* Paper No. 1.3. Inst. Chem. Eng., London.
62. Fisons Ltd.—Fertilizer Div., Levington Res. Station, Ipswich, U.K., Personal communication from J. A. Storrow, 1969.
63. Fleming, R. J., The spoutability of particulate solids in air. M.A.Sc. thesis,Univ. of Toronto, Canada, 1966.
64. Frigoscandia, Sweden, "FloFREEZE" Manual No. 2-30-2, 681201.
65. F. W. Horner Ltd., Montreal, Canada. Personal communication from M. Frechette, 1968.
66. Gabor, J. D., Wall-to-bed heat transfer in fluidized and packed beds. *Chem. Eng. Progr. Symp. Ser.* **66,** No. 105, 76 (1970).
67. Galkin, O. A., Romankov, P. G., Tanganov, I. N., and Frolov, V. F., Investigation of statistical characteristics of solids flow in a spouted bed. *Teor. Osn. Khim. Tekhnol.* **2,** 884 (1968).
68. Gay, E., Nelson, G. L., and Clary, B. L., Air flow requirements and bed turnover time for a spouted bed peanut drier. *Meeting of Amer. Soc. Agr. Eng., Minneapolis, 1970,* Paper No. 70-309. ASAE, St. Joseph, Michigan.
68a. Gay, E., Particle and fluid transport characteristics of spouted beds for whole Spanish peanuts. M.S. thesis, Oklahoma State Univ., Stillwater, 1970.
69. Gibson, A., "Hydraulics and Its Applications," 5th ed. Constable Press, London, 1952. Quoted by Perry [177, p. 5–32].
70. Gishler, P. E., and Mathur, K. B., Method of contacting solid particles with fluids. U.S. Patent No. 2,786,280 to Nat. Res. Council of Can., 1957 (filed 1954). Brit. Patent No. 801,315.
71. Gelperin, N. I., Ainshtein, V. G., Gelperin, E. N., and L'vova, S. D., Hydrodynamic properties of fluidized granular materials in conical and conical-cylindrical sets. *Khim. Tekhnol. Top. Masel* **5,** No. 8, 51 (1960).
72. Gelperin, N. I., Ainshtein, V. G., and Timokhova, L. P., Hydrodynamic features of fluidization of granular materials in conical sets. *Khim. Mashinostr. (Moscow)* **No. 4,** 12 (1961).
73. Gelperin, N. I., and Fraiman, R. S., Study of heat release from the conical surface to a fluidized bed. *Khim. Prom. (Moscow)* **No. 11,** 6 (1963).
74. Gelperin, N. I., Vainberg, Y. P., and Ainshtein, V. G., Continuous granulation of powders of medicinal preparations in a fluidized bed. *Med. Ind. U.S.S.R.* **No. 10,** 27 (1965).
75. Ghosh, B., A study on the spouted bed—A theoretical analysis. *Indian Chem. Engr.* **7,** 16, (1965).
76. Ghosh, B., and Osberg, G. L., Heat transfer in water spouted beds. *Can. J. Chem. Eng.* **37,** 205 (1959).
77. Goldberger, W. M., and Nack, H., Novel uses of fluidized beds in chemical processing. *Battelle Tech. Rev.* **13,** 3 (1964).
78. Goltsiker, A. D., Doctoral dissertation, Lensovet Technol. Inst., Leningrad, 1967. Quoted by Romankov and Rashkovskaya [201, Chapter 1].
79. Goltsiker, A., Rashkovskaya, N. B., and Romankov, P. G., The mechanism of commencement of boiling in conical sets. *Zh. Prikl. Khim. (Leningrad)* **37,** 1030 (1964).
80. Golubev, L. G., and Nikolaev, A. M., Drying of granular substances in air spouting equipment. *Publ. Kazan Chem. Technol. Inst. (Kirov)* **32,** 137 (1964).
81. Golubkovich, A. V., Kondukov, N. B., and Vorob'ev, Kh. S., Some hydrodynamic features of spouting-pulsating fluidization of granular material in conical apparatus. *Khim. Prom. (Moscow)* **43,** 526 (1967).

82. Golubkovich, A. V., Kondukov, N. B., and Vorob'ev, Kh. S., Structure and stability of spouted bed in cone-bottomed columns. *Teor. Osn. Khim. Tekhnol.* **2**, 879 (1968).

83. Gorshtein, A. E., and Soroko, V. E., Piezoelectric method of studying a suspended layer. *Izv. Vyssh. Ucheb. Zaved. Khim. Khim. Tekhnol.* **4**, No. 1, 137 (1964).

84. Gorshtein, A. E., and Mukhlenov, I. P., Critical speed of gas corresponding to the beginning of spouting. *Zh. Prikl. Khim.* (*Leningrad*) **37**, 1887 (1964).

85. Gorshtein, A. E., and Mukhlenov, I. P., On the mechanics of formation of spouting beds. *Zh. Prikl. Khim.* (*Leningrad*) **38**, 16 (1965).

86. Gorshtein, A. E., and Mukhlenov, I. P., The movement of solid material in the spouting bed. *Zh. Prikl. Khim.* (*Leningrad*) **40**, 2469 (1967).

87. Green, D. J., Dynamics of a spouted liquid reactor. M.Sc. thesis, Univ. of New Brunswick, Fredericton, Can., 1967.

88. Haji-Zainali, M., Solids size reduction in a spouted bed. B.A.Sc. thesis, Univ. of Brit. Columbia, Vancouver, Can., 1970.

89. Happel, J., Pressure drop due to vapor flow through moving beds. *Ind. Eng. Chem.* **41**, 1161 (1949).

90. Heertjes, P. M., De Nie, L. H., and Verloop, J., The manufacture of Portland cement clinker in a spouting bed. *Powder Technol.* **4**, 269 (1970–1971).

91. Heiser, A. L., Lowenthal, W., and Singiser, R. E., Method and apparatus for coating particles. U.S. Patent No. 3,112,220 to Abbott Lab., Chicago, Illinois, 1963 (filed 1960).

92. Higbie, R., The rate of absorption of a pure gas into a still liquid during short periods of exposure. *Trans. A.I.Ch.E.* **31**, 365 (1935).

93. Huey, L. F., The coating of pelletized fish food in a spouted bed. B.A.Sc. thesis, Univ. of Brit. Columbia, Vancouver, Can., 1971.

94. Hunt, C. H., and Brennan, D., Estimation of spout diameter in a spouted bed. *Aust. Chem. Eng.* **5**, 9 (1965).

95. I.C.I. Fibres Ltd., Harrogate, U.K., Personal communication, 1969.

96. Indian Explosives Ltd., Gomia, Bihar. Work carried out in 1966 under supervision of K. B. Mathur.

97. Jakob, M., "Heat Transfer," Vol. 1, Chapters 13 and 20. Wiley, New York, 1949.

98. Johnston, T. R., Robinson, C. W., and Epstein, N., A spouted mixer-settler. *Can. J. Chem. Eng.* **39**, 1 (1961).

99. Kazakova, E. A., The application of fluidized bed for intensification of prilling nitrogen fertilizers. *Int. Congr. Chem. Eng.* (*CHISA*), *3rd, Prague, September 1969*, Paper No. C3,11, Czechoslovak Society for Industrial Chemistry.

100. Klassen, J., and Gishler, P. E., Heat transfer from column wall to bed in spouted, fluidized and packed systems. *Can. J. Chem. Eng.* **36**, 12 (1958).

101. Klimenko, Y. G., and Rabinovich, M. I., A study of the process of drying crystal hydrates of chloride of manganese in a spouting bed. Collection of papers on "Heat and Mass Exchange," p. 118. Nauk. Dumka, Ukr. SSR Acad. of Sci., Kiev, 1968.

102. Klimenko, Yu. G., Karpenko, V. G., and Rabinovich, M. I., Heat exchange between the spouting bed and the surface of a spherical probe element. "Heat Physics and Heat Technology," No. 15, p. 81. Ukr. SSR Acad. of Sci., Kiev, 1969.

102a. Kotzo, C. J., Solids flow in a spouted bed—Computation work. B.A.Sc. thesis, Univ. of Brit. Columbia, Vancouver, Can., 1970.

103. Koyanagi, M., The design, construction and determination of the properties of a spouted bed. B.A.Sc. thesis, Univ. of Brit. Columbia, Vancouver, Can., 1955.

104. Kugo, M., Watanabe, N., Uemaki, O., and Shibata, T., Drying of wheat by spouting bed. *Bull. Hokkaido Univ. Sapporo, Jap.* **39**, 95 (1965).

105. Kutsakova, V. E., Romankov, P. G., and Rashkovskaya, N. B., Some kinetic relationships of drying in fluidized and spouted beds—Pt. I. *Zh. Prikl. Khim.* (*Leningrad*) **36**, 2217 (1963).

106. Kutsakova, V. E., Romankov, P. G., and Rashkovskaya, N. B., Some kinetic relationships of drying in fluidized and spouted beds—Pt. II, *Zh. Prikl. Khim. (Leningrad)* **37**, 1972 (1964).

107. Kutsakova, V. E., Romankov, P. G., and Rashkovskaya, N. B., Some kinetic relationships of drying in fluidized and spouted beds—Pt. III, *Zh. Prikl. Khim. (Leningrad)* **37**, 2223 (1964).

108. Kunii, D., and Smith, J. M., Heat transfer characteristics of porous rocks. *A.I.Ch.E. J.* **6**, 71 (1960).

109. Kunii, D., and Okayashu, S., *Kagaku Kogaku* **31**, 699 (1967). Quoted by Uemaki *et al.* [237].

110. Kunii, D., and Levenspiel, O., "Fluidization Engineering." Wiley, New York, 1969.

111. Lama, R. F., Pressure drop in spouted beds. M.Sc. thesis, Univ. of Ottawa, Ottawa, Can., 1957.

112. Lang, E. W., Smith, H. G., and Bordenca, C., Carbonization of agglomerating coals in a fluidized bed. *Ind. Eng. Chem.* **49**, 355 (1957).

113. Lefroy, G. A., The mechanics of spouted beds. Ph.D. thesis, Univ. of Cambridge, Cambridge, England, 1966.

✔ 114. Lefroy, G. A., and Davidson, J. F., The mechanics of spouted beds. *Trans. Inst. Chem. Eng.* **47**, T120 (1969).

115. Lefroy, G. A., and Davidson, J. F., *Chem. Eng. (London)* **No. 233**, CE 390 (1969).

116. Leva, M., Weintraub, M., Grummer, M., Pollchik, M., and Storch, H. H., Fluid flow through packed and fluidized systems. *U.S. Bur. Mines Bull.* **504** (1951).

117. Leva, M., "Fluidization." McGraw-Hill, New York, 1959.

118. Levenspiel, O., "Chemical Reaction Engineering." Wiley, New York, 1962.

119. Levenspiel, O., Mixed models to represent flow of fluids through vessels. *Can. J. Chem. Eng.* **40**, 135 (1962).

120. Levenspiel, O., and Bischoff, K. B., Patterns of flow in chemical process vessels. *Advan. Chem. Eng.* **4**, 95 (1963).

121. Lim, C. J. Unpublished work. Univ. of Brit. Columbia, Vancouver, Can., 1972.

122. Littman, H., and Sliva, D. E., Gas–particle heat transfer coefficients in packed beds at low Reynolds number. "Heat Transfer 1970, Paris–Versailles," Elsevier, Amsterdam, 1971. **7**, CT 1.4.

123. Lutz, W. A., Fertilizer manufacture. U.S. Patent No. 2,600,253 to Dorr Co., Stamford, Connecticut, 1952 (filed 1949).

124. Madonna, L. A., and Lama, R. F., Pressure drop in spouted beds. *Ind. Eng. Chem.* **52**, 169 (1960).

125. Madonna, L. A., Lama, R. F., and Brisson, W. L., Solids–air jets. *Brit. Chem. Eng.* **6**, 524 (1961).

126. Malek, M. A., Heat transfer in spouted beds. Ph.D. thesis, Univ. of Ottawa, Ottawa, Can., 1963.

127. Malek, M. A., Madonna, L. A., and Lu, B. C. Y., Estimation of spout diameter in a spouted bed. *Ind. Eng. Chem. Process Des. Develop.* **2**. 30 (1963).

128. Malek, M. A., and Lu, B. C. Y., Heat transfer in spouted beds. *Can. J. Chem. Eng.* **42**, 14 (1964).

129. Malek, M. A., and Lu, B. C. Y., Pressure drop and spoutable bed height in spouted beds. *Ind. Eng. Chem. Process Des. Develop.* **4**, 123 (1965).

130. Malek, M. A., and Walsh, T. H., The treatment of coal for coking by the spouted bed process. Rep. No. FMP 66/54-SP. Dept. Mines and Tech. Surveys, Ottawa, Can., 1966.

131. Mamuro, T., and Hattori, H., Flow pattern of fluid in spouted beds. *J. Chem. Eng. Jap.* **1**, 1 (1968); Correction. *J. Chem. Eng. Jap.* **3**, 119 (1970).

132. Mann, U., and Crosby, E. J., Modeling circulation of solids in spouted beds. *Ind. Eng. Chem. Process Des. Develop.* **11**, 314 (1972).

133. Mann, U., and Crosby, E. J., Cycle time distribution in circulating systems. *Chem. Eng. Sci.* **28**, 623 (1973).

134. Manurung, F., Studies in the spouted bed technique with particular reference to low temperature coal carbonization. Ph.D. thesis, Univ. of New South Wales, Kensington, Australia, 1964.

135. Massimilla, L., Volpicelli, G., and Raso, G., A study on pulsing gas fluidization of beds of particles. *Chem. Eng. Progr. Symp. Ser.* **62.** 63 (1966).

136. Matheson, G. I., Herbst, W. A., and Holt, P. H., Characteristics of fluid–solid systems. *Ind. Eng. Chem.* **41.** 1099 (1949).

137. Mathur, K. B., and Gishler, P. E., A technique for contacting gases with coarse solid particles. *A.I.Ch.E. J.* **1,** 157 (1955).

138. Mathur, K. B., and Gishler, P. E., A study of the application of the spouted bed technique to wheat drying. *J. Appl. Chem.* **5.** 624 (1955).

139. Mathur, K. B., and Gishler, P. E., Mass transfer in a spouted bed. *C.I.C. Annu. Conf., Quebec, 1955,* unpublished work.

140. Mathur, K. B., Spouted beds. *In* "Fluidization" (J. F. Davidson and D. Harrison, eds.), Chapter 17. Academic Press, New York, 1971.

141. Mathur, K. B., and Epstein, N., Developments in spouted bed technology. *Can. J. Chem. Eng.* **52,** 129 (1974); *Energ. Atomtech.* (*Budapest*) (to be published).

142. Mathur, K. B., and Epstein, N., Dynamics of spouted beds. *Advan. Chem. Eng.* **9,** 111–191 (1974).

143. Mathur, K. B., A new technique for grinding of particulate solids. Patent proposal, 1972. Prelim. work reported in B.A.Sc. theses by C. L. Connaghan, 1970 and D. G. Miller, 1971, Univ. of Brit. Columbia, Vancouver, Can.

144. Mathur, K. B., and Lim, C. J., Vapour phase chemical reaction in spouted beds—A theoretical model. *Chem. Eng. Sci.* **29,** 789 (1974).

145. Matsen, J. M., Void fraction variation in the spouted bed annulus. *Ind. Eng. Chem. Process Des. Develop.* **7,** 159 (1968).

145a. Matsen, J. M., Personal communication, 1968.

146. May, W. G., Fluidized-bed reactor studies. *Chem. Eng. Progr.* **55.** No. 12, 49 (1959).

147. McEwen, E., Simmonds, W. H. C., and Ward, G. T., The drying of wheat grain, Pt. III: Interpretation in terms of its biological structure. *Trans. Inst. Chem. Eng.* **32,** 115 (1954).

148. McNab, G. S., Prediction of spout diameter. *Brit. Chem. Eng. Proc. Tech.* **17.** 532 (1972).

149. Meisen, A., Univ. of Brit. Columbia, Vancouver. Can. Personal communication, 1971.

150. Meisen, A., and Mathur, K. B., The spouted bed aerosol collector: A novel device for separating small particles from gases. Multi-phase Flow Systems. *Inst. Chem. Eng. Symp. Ser. No. 38* (*1974*), Paper K3. Prelim. work reported in B.A.Sc. thesis by R. Mia, Univ. of Brit. Columbia, Vancouver, Can., 1973.

151. Metheny, D. E., and Vance, S. W., Particle growth in fluidized bed dryers. *Chem. Eng. Progr.* **58.** 45 (1962).

152. Mikhailik, V. D., The pattern of change of spout diameter in a spouting bed. Collected works on "Research on Heat and Mass Transfer in Technological Processes," p. 37. Nauka i Tekhnika BSSR, Minsk, 1966.

153. Mikhailik, V. D., and Antanishin, M. V., The speed of particles and voidage in the core of the spouting bed. *Vesti Akad. Nauk. BSSR Minsk Ser. Fiz. Takhn. Nauk* No. 3, 81 (1967).

154. Minchev, A. D., Romankov, P. G., and Rashkovskaya, N. B., Investigation of drying of pastes in spouted beds of inert materials. *Zh. Prikl. Khim.* (*Leningrad*) **41,** 1249 (1968).

155. Minchev, A. D., Romankov, P. G., and Rashkovskaya, N. B., Determination of the optimum conditions for drying paste-like materials in a spouting bed of inert solids. *Zh. Prikl. Khim.* (*Leningrad*) **42.** 2150 (1969).

156. Mitev, D. T. Doctoral dissertation Leningrad Inst. of Technol., Leningrad, 1967. Quoted by Romankov and Rashkovskaya [201, Chapter 1].

157. Miyahara, Y., and Ozawa, T., *J. Jap. Petrol. Inst.* **3,** 317 (1960). Quoted by Uemaki *et al.* [237].
158. Mukhlenov, I. P., and Gorshtein, A. E., Hydraulic resistance of the suspended layer in grateless conical sets. *Zh. Prikl. Khim. (Leningrad)* **37.** 609 (1964).
159. Mukhlenov, I. P., and Gorshtein, A. E., Investigation of a spouting bed. *Khim. Prom. (Moscow)* **41,** 443 (1965).
160. Mukhlenov, I. P., and Gorshtein, A. E., Hydrodynamics of reactors with a spouting bed of granular materials. *Vses. Konf. Khim. Reactrom Novosibirsk.* **3.** 553 (1965).
161. Murray, F. E., Prakash, C. B., and Mathur, K. B., Process and apparatus for elimination of hydrogen sulfide emissions during calcining and drying of wet calcium carbonate dust. Patent proposal, 1973. Univ. of Brit. Columbia, Vancouver, Can.
162. Nack, H., Fluidized-bed processing of nuts. *Mfg. Confect.* **46** (1966).
163. Nautamix Co., Haarlem, Holland, Brochures entitled "Vometec N.V." and "Nauta Vometec Fluid Bed Reactor"; also G. T. King, Mixing with reaction. *Chem. Process Eng.* **48.** 72 (1967).
164. Nelson, G. L., and Gay, E., Spouted bed fluid and particle transport processes for coarse biological materials. *Meeting Amer. Soc. Agr. Eng., Lafayette, 1969,* Paper No. 69-371. ASAE, St. Joseph, Michigan.
165. Németh, J., Pallai, I., and Baticz, S., Continuous adsorption equipment. *Proc. Conf. Some Aspects Phys. Chem., Budapest, 1966,* p. 259. Contribution from Res. Inst. for Tech. Chem., Hung. Acad. of Sci., Budapest.
166. Németh, J., and Pallai, I., Spouted bed technique and its application. *Magy. Kem. Lapja.* **25.** 74 (1970).
167. Nichols, F. P., Improvements in and relating to the production of granular compositions such as fertilizers. Brit. Patent No. 1,039,177 to ICI Ltd., London, 1966 (filed 1963).
168. Nikolaev, A. M., and Golubev, L. G., Basic hydrodynamic characteristics of the spouting bed. *Izv. Vyssh. Ucheb. Zaved. Khim. Khim. Tekhnol.* **7.** 855 (1964).
169. Orr, C., "Particulate Technology," Chapter 5. Macmillan, New York, 1966.
170. Osberg, G. L., Locating fluidized solids bed level in a reactor—Hot wire method. *Ind. Eng. Chem.* **43.** 1871 (1951).
171. Pallai, I. V. E., Research on the fluid mechanics of spouted beds. Dissertation Abstract, Res. Inst. for Tech. Chem., Hung. Acad. of Sci., Budapest, 1971.
172. Pallai, E., Examination of heat transfer in particle aggregations. *Proc. Int. Conf. Appl. Chem., Veszprem, 2nd, 1971,* p. 115. Contribution from Res. Inst. for Tech. Chem., Hung. Acad. of Sci., Budapest.
173. Pallai, I., and Németh, J., Analysis of flow forms in a spouted bed apparatus by the so-called phase diagram. *Int. Congr. Chem. Eng. (CHISA), 3rd, Prague, September 1969,* Paper No. C2.4. Czechoslovak Society for Industrial Chemistry.
174. Pallai, E., and Németh, J., Measurement and regulation of particle residence time in spouted bed driers. *Int. Conf. Drying, 3rd, Budapest, September 1971,* Paper No. D.7. Contribution from Res. Inst. for Tech. Chem., Hung. Acad. of Sci., Budapest.
175. Pallai, I., and Németh, J., Residence time distribution in spouting beds. *Int. Congr. Chem. Eng. (CHISA), 4th, Prague, September 1972,* Paper No. C3.11. Czechoslovak Society for Industrial Chemistry.
176. Peltzman, A., and Pfeffer, R., The effect of a small number of inert particles on the local and overall mass transfer rates from a single sphere at low Reynolds numbers. *Chem. Eng. Progr. Symp. Ser.* **63.** No. 77, 49 (1967).
177. Perry, J. H., "Chemical Engineer's Handbook," 4th ed., McGraw-Hill, New York, 1963.
178. Peterson, W. S., Spouted bed drier. *Can. J. Chem. Eng.* **40.** 226 (1962).

179. Peterson, W. S., Multiple spouted bed. Can. Patent No. 739,660 to Nat. Res. Council of Can., 1966 (filed 1963). Prelim. work reported in an internal Nat. Res. Council report by Peterson, 1961.

180. Peterson, W. S., Chem. Div., Nat. Res. Council of Can., Ottawa. Personal communications, 1969–1972.

181. Pitt, G. J., The kinetics of the evolution of volatile products from coal. Fuel **41**, 267 (1962).

182. Pomortseva, A. A., and Baskakov, A. P., Hydrodynamics and heat transfer in fluidized beds of fine grained materials with local spouting zones. *Khim Tekhnol. Topl. Masel* **15**, 34 (1970).

183. Popov, V. A., Romankov, P. G., and Rashkovskaya, N. B., Evenness of drying of some granular polymer materials in a spouting bed. *Zh. Prikl. Khim.* (*Leningrad*) **43**, 2561 (1970).

184. Prados, J. W., and Scott, J. L., Mathematical model for predicting coated-particle behaviour. *Nucl. Appl.* **2**, 402 (1966).

185. Quick Frozen Foods. Fluidized unit ups capacity, cuts space for vegetable, fruit freezers. *Quick Frozen Foods* **29**(5), 280 (December 1966).

186. Quinlan, M. J., and Ratcliffe, J. S., Consequential effects of air drying wheat—spouted bed design and operation. *Mech. Chem. Eng. Trans.* **6**, 19 (1970).

187. Quinlan, M. J., and Ratcliffe, J. S., A design equation for the low temperature carbonization of coal under spouted bed conditions. *Mech. Chem. Eng. Trans.* **7**, 1 (1971).

188. Rao, M. V. R., Nomograph determines fluid velocity needed for spouting. *Chem. Eng.* (*New York*) **74**, 122 (1967).

189. Rashkovskaya, N. B., Investigation of hydrodynamics and dehydration process in slot apparatus with a whirlwind bed. *Int. Congr. Chem. Eng.* (*CHISA*), *4th, Prague, September 1972*, Paper No. E5.9. Czechoslovak Society for Industrial Chemistry.

190. Ratcliffe, J. S., and Rigby, G. R., Low temperature carbonization of coal in a spouted bed—Prediction of exit char volatile matter. *Mech. Chem. Eng. Trans.* **5**, 1 (1969).

191. Reddy, K. V. S., Studies in the spouting of mixed particle size beds. M.A.Sc. thesis, Univ. of Brit. Columbia, Vancouver, Can., 1963.

192. Reddy, K. V. S., Fleming, R. J., and Smith, J. W., Maximum spoutable bed depths of mixed particle-size beds. *Can. J. Chem. Eng.* **46**, 329 (1968).

193. Reger, E. O., Romankov, P. G., and Rashkovskaya, N. B., Drying of paste-like materials on inert bodies in a spouting bed. *Zh. Prikl. Khim.* (*Leningrad*) **40**, 2276 (1967); also, collected papers on "Processes of Chemical Technology," p. 349. Nauka, 1965, quoted by Romanakov and Rashkovskaya [201].

194. Richardson, J. F., and Zaki, W. N., Sedimentation and fluidization. *Trans. Inst. Chem. Eng.* **32**, 35 (1954).

195. Richardson, J. F., and Mitson, A. E., *Trans. Inst. Chem. Eng.* **36**, 270 (1958).

196. Robinson, C. E., Improvement in furnaces for roasting ore. U.S. Patent No. 212,508; 1879.

197. Roblee, L. H. S., Baird, R. M., and Tierney, J. W., Radial porosity variations in packed beds. *A.I.Ch.E. J.* **4**, 461 (1958).

198. Rockwell, W. C., Lowe, E., Walker, H. G., and Morgan, A. I., Hot air grain popping. *Feedlot* (November 1968).

199. Romankov, P. G., Drying. *In* "Fluidization" (J. F. Davidson and D. Harrison, eds.), Chapter 12. Academic Press, New York, 1971.

200. Romankov, P. G., Rashkovskaya, N. B., and Berezovskaya, Z. A., New methods of drying paste-like pigments. *Lakokrasoch. Mat. Ikh Primen.* No. 3, p. 71 (1960).

201. Romankov, P. G., and Rashkovskaya, N. B., "Drying in a Suspended State," 2nd ed., in Russian. Chem. Publ. House, Leningrad Branch, 1968.

202. Romankov, P. G., Rashkovskaya, N. B., Goltsiker, A. D., and Seballo, V. A., A study of the structure of spouting beds. *Khim. Prom. (Moscow)* **46**, 372 (1970); Engl. transl. *Heat Transfer Sov. Res.* **3**, 133 (1971).

203. Rooney, N. M., and Harrison, D., Spouted beds of fine particles. *Powder Technol.* **9**, 227 (1974).

204. Rowe, P. N., and Claxton, K. T., Heat and mass transfer from a single sphere to fluid flowing through an array. *Trans. Inst. Chem. Eng.* **43**. T321 (1965).

205. Russell, C. C., Carbonization. *In* "Encyclopedia of Chemical Technology" (R. E. Kirk and D. F. Othmer, eds.), Vol. 4. Wiley (Interscience), New York, 1964.

206. Sarkits, V. B., Heat transfer from suspended beds of granular materials to heat transfer surfaces. Dissertation, Leningrad, 1959. Quoted by Zabrodsky [260, p. 278].

207. Sazhin, B. S., An investigation of the process of drying free flowing materials in spouting driers and development of practical drying equipment. *Mat. Soveshch. Teplo i Massobmenu (Dokl.) Minsk,* p. 169 (1961).

208. Schlünder, E. U., Heat transfer in moving packings of spherical particles in contact for short periods. *Chem. Ing. Tech.* **43**, 651 (1971).

209. Schofield, B., Solids size reduction in a spouted bed. B.A.Sc. thesis, Univ. of Brit. Columbia, Vancouver, Can., 1969.

210. Sevilla, E., and Pinder, K. L. Personal communication from K. L. Pinder, Univ. of Brit. Columbia, Vancouver, Can.; work done at Univ. of Havana, Cuba, 1970.

211. Shakhova, N. A., Yevdokimov, B. G., and Ragozina, N. M., An investigation of a multi-compartment fluid-bed granulator. *Proc. Tech. Int. London* **17**, 946 (1972).

212. Shakhova, N. A., Tikhonov, I. D., and Ragozina, N. M., Apparatus for granulating urea in a fluidized bed. *Chem. Petrol. Eng.* Nos. 9 and 10, p. 785 (1970); translated from *Khim. i. Neft. Mashinostr.* No. 9, p. 41 (1970).

213. Shigeo, N., Particle behaviour in spouted beds. Bachelor's thesis, Hokkaido Univ., Sapporo, Japan, 1965.

214. Shirai, T., "Fluidized Beds" (in Japanese). Kagaku-Gijutsusha, Kanazawa, 1958. Quoted by Uemaki *et al.* [237].

215. Shreve, R. N., "The Chemical Process Industries," 2nd ed. McGraw-Hill, New York, 1956.

216. Singiser, R. E., and Lowenthal, W., Enteric film coats by the air-suspension coating technique. *J. Pharm. Sci.* **50**. 168 (1961).

217. Singiser, R. E., Heiser, A. L., and Prillig, E. B., Air-suspension tablet coating. *Chem. Eng. Progr.* **62**, No. 6, 107 (1966).

218. Sissom, L. E., and Jackson, T. W., Heat exchange in fluid–dense particle moving beds. *J. Heat Transfer* **89**, 1 (1967).

219. Smith, J. W., and Reddy, K. V. S., Spouting of mixed particle-size beds. *Can. J. Chem. Eng.* **42**, 206 (1964).

220. Smith, W. A., Minimum fluid velocity in spouting beds. B.A.Sc. thesis, Univ. of Brit. Columbia, Vancouver, Can., 1969.

221. Suciu, G. C., and Patrascu, M., Hydrodynamic characteristics of the spouted bed. *Chim. Ind. Genie Chim.* **104**, 1304 (1971).

222. Sutherland, J. P., The measurement of pressure drop across a gas fluidized bed. *Chem. Eng. Sci.* **19**, 839 (1964).

223. Syromyatnikov, N. I., Results of tests on furnaces of the fluidized, suspended and spouting types. *Za Ekonom Topl.* No. 2, p. 17 (1951).

224. Takahashi, H., and Yanai, H., Flow profile and void fraction of granular solids in a moving bed. *Powder Technol.* **7**, 205 (1973).

225. Thoenes, D., and Kramers, H., Mass transfer from spheres in various regular packings to a flowing fluid. *Chem. Eng. Sci.* **8**, 271 (1958).

226. Thomas, S., Optimization of a spouted liquid reactor. M.Sc. thesis, Univ. of New Brunswick, Fredericton, Can., 1970.

227. Thorley, B., Mathur, K. B., Klassen, J., and Gishler, P. E., The effect of design variables on flow characteristics in a spouted bed. Rep. Nat. Res. Council of Can., Ottawa, 1955.

228. Thorley, B., Saunby, J. B., Mathur, K. B., and Osberg, G. L., An analysis of air and solid flow in a spouted wheat bed. *Can. J. Chem. Eng.* **37**, 184 (1959).

229. Tsvik, M. Z., Nabiev, M. N., Rizaev, N. U., Merenkov, K. V., and Vyzgo, V. S., The minimum velocity for internal spouting in the combined process for producing granulated fertilizers. *Uzb. Khim. Zh.* **10**, No. 6, 3 (1966).

230. Tsvik, M. Z., Nabiev, M. N., Rizaev, N. U., Merenkov, K. V., and Vyzgo, V. S., The velocity for external spouting in the combined process for production of granulated fertilizers. *Uzb. Khim. Zh.* **11**, No. 2, 50 (1967).

231. Tsvik, M. Z., Nabiev, M. N., Rizaev, N. U., and Merenkov, K. V., Angular value of a spouting core. *Uzb. Khim. Zh.* **11**, No. 4, 64 (1967).

232. Tsvik, M. Z., Nabiev, M. N., Risaev, N. U., and Merenkov, K. V., Process of mass transfer in the combined process for production of complex granulated fertilizer. *Uzb. Khim. Zh.* **11**, No. 5, 69 (1967).

233. Turner, G. A., The thermal history of a granule in a rotary cooler. *Can. J. Chem. Eng.* **44**, 13 (1966).

234. Uemaki, O., The behaviour of solid particles in spouted beds and the application of spouted bed technique to thermal cracking of heavy oil. Ph.D. thesis, Hokkaido Univ., Sapporo, Japan, 1968 (in Japanese).

235. Uemaki, O., and Kugo, M., Heat transfer in spouted beds. *Kagaku Kogaku* **31**, 348 (1967).

236. Uemaki, O., and Kugo, M., Mass transfer in spouted beds. *Kagaku Kogaku* **32**, 895 (1968).

237. Uemaki, O., Fugikawa, M., and Kugo, M., Pyrolysis of petroleum (crude oil, heavy oil and naphtha) for production of ethylene and propylene in an externally heated spouted bed. *Kogyo Kagaku Zasshi* **73**, 453 (1970).

238. Uemaki, O., Fugikawa, M., and Kugo, M., Pyrolytic cracking of petroleum fractions (kerosene, heavy oil and crude oil) by using a two-spouted bed reactor system. *Kogyo Kagaku Zasshi* **74**, 933 (1971).

239. Uemaki, O., and Mathur, K. B., Granulation of ammonium sulphate fertilizer in a spouted bed. Unpublished work, Univ. of Brit. Columbia, Vancouver, Can., 1972; also, K. M. Ma. B.A.Sc. thesis, Univ. of Brit. Columbia, Vancouver, Can., 1973.

240. Vainberg, Yu. P., Granulation of medicinal preparations in multi-stage continuously-operated conical fluidized beds. Dissertation Abstract, Moscow Inst. of Fine Chem. Technol., 1968 (in Russian).

241. Vainberg, Yu. P., Gelperin, N. I., and Ainshtein, V. G., On the unique behaviour of granular substances during fluidization in conical sets. Collection of papers in "Processes and Equipment of Chemical Technology" (N. I. Gelperin, ed.), p. 22. Ministry of Higher Education, Moscow, 1967.

242. Vanecek, V. M., Markwart, M., and Drbohlav, R., "Fluidized Bed Drying." Leonard Hill, London, 1966.

243. van der Merwe, D. F., and Gauvin, W. H., Pressure drag measurements for turbulent air flow through a packed bed. *A.I.Ch.E. J.* **17**, 402 (1971).

244. van Velzen, D., Flamm, H. J., and Langenkamp, H., Gas flows in spouting beds. Dragon Proj. Rep. No. 785. Euratom, Ispra, Italy, 1972.

245. Vavilov, N. S., Tsylev, L. M., and Chao, C.-C., Reduction of iron ores in a spouting fluidized bed. *Russ. Met. Fuels* No. 1, 23 (1962).

246. Vavilov, N. S., and Chao, C.-C., Laboratory installation for the investigation of the physico-chemical and gas-flow processes in a spouted bed. *Tr. Inst. Met. Amad. Nauk SSSR im. A. A. Baikova* No. 12, p. 41 (1962).

247. Vavilov, N. S., Melent'ev, P. N., Tsylev, L. M., and Chao, C.-C., Spouted bed reduction of iron ore and ore–carbon granules in a stream of natural gas. *Russ. Met.* No. 1, 7 (1967).

248. Volkov, A. I., Romankov, P. G., Frolov, V. F., Taganov, I. N., and Galkin, O. A., Statistical characteristics of gas motion in the slot of an apparatus with a spouted bed. *Zh. Prikl. Khim. (Leningrad)* **43**, 1079 (1970).

249. Volpicelli, G., Spouted bed catalytic reactors with continuous and pulsing feed. *Quad. Ing. Chim. Ital.* **1**, 37 (1965).

✓250. Volpicelli, G., and Raso, G., Flow states of granular beds with through-flow of gas in a continuous jet. *Atti Accad. Naz. Lincei Rend. Cl. Sci. Fis. Mat. Natur.* **35**, 331 (1963).

251. Volpicelli, G., Raso, G., and Saccone, L., Gas–solid systems with pulsating feed. *Chim. Ind. (Milan)* **45**, 1362 (1963).

✓ 252. Volpicelli, G., Raso, G., and Massimilla, L., Gas and solid flow in bidimensional spouted beds. *Proc. Eindhoven Fluidizn. Symp., 1967*, p. 123. Netherlands Univ. Press, Amsterdam.

253. Vuković, D. V., Zdanski, F. K., and Grbavcić, Z., Effect of annular and nozzle flow of fluid on the behavior of spouted-fluidized bed. *Int. Congr. Chem. Eng. (CHISA), 4th, Prague, September 1972*, Paper No. C3.8. Czechoslovak Society for Industrial Chemistry.

254. Vuković, D. V., Zdanski, F. K., and Vunjak, G. V., The three-phase spouted bed—A new system in chemical engineering processing. *A.I.Ch.E. Meeting, Detroit, June 1973*, Paper No. 11d. A.I.Ch.E., New York.

255. Vyzgo, V. S., Pavlova, A. I., and Nabiev, M. N., The possibility of intensifying the process of preparation of fertilizers by using fluidization. *Uz. Khim. Zh.* **9**, No. 4, 5 (1965).

✓ 256. Wan-Fyong, F., Romankov, P. G., and Rashkovskaya, N. B., Research on the hydrodynamics of the spouting bed. *Zh. Prikl. Khim. (Leningrad)* **42**, 609 (1969).

257. Webster, G. W., Comparison of the behaviour of conical and cylindrical spouted beds. B.A.Sc. thesis, Univ. of Brit. Columbia, Vancouver, Can., 1972.

258. Wen, C. Y., and Yu, Y. H., A generalized method for predicting the minimum fluidization velocity. *A.I.Ch.E. J.* **12**, 610 (1966).

259. Wolfram, C. F., Spout shape in spouted beds. B.A.Sc. thesis, Univ. of Brit. Columbia, Vancouver, Can., 1972.

260. Zabrodsky, S. S., "Hydrodynamics and Heat Transfer in Fluidized Beds." MIT Press, Cambridge, Massachusetts, 1966.

261. Zabrodsky, S. S., and Mikhailik, V. D., The heat exchange of the spouting bed with a submerged heating surface. Collected papers on "Intensification of Transfer of Heat and Mass in Drying and Thermal Processes," p. 130. Nauka i Tekhnika BSSR, Minsk, 1967.

262. Zanker, E. A., Nomograph for calculating the superficial fluid velocity in spouted beds. *Chem. Eng. (London)* **No. 242**, CE 351 (1970).

263. Zenz, F. A., and Othmer, D. F., "Fluidization and Fluid-particle Systems." Van Nostrand-Reinhold, Princeton, New Jersey, 1960.

The following group of papers in Japanese by A. Yokogawa and co-workers escaped our notice. These were brought to our attention by Prof. V. D. Vuković of Belgrade University, but only after the manuscript of this book was nearly complete.

264. Nikajima, Y., and Yokogawa, A., Experiments on fluidization of fluidized and spouted bed. *Hitachi Zosen Tech. Rev.* **38**, 109 (1963).

265. Yokogawa, A., Suzuki, N., Ogino, E., and Yoshii, N., Spout diameter and spouting bed height in the spouted bed. *Trans. Japan Soc. Mech. Eng.* **35**, 1903 (1969).

266. Yokogawa, A., Ogino, E., and Yoshii, N., On the conditions of spouting granular solid beds. *Trans. Japan Soc. Mech. Eng.* **36**, 365 (1970).
267. Yokogawa, A., Ogino, E., and Yoshii, N., Pressure drop at the start of and during spouting in the spouted bed. *Trans. Japan Soc. Mech. Eng.* **36**, 375 (1970).
268. Yokogawa, A., Ogino, E., and Yoshii, N., Flow pattern of particles in the annulus of the spouted bed. *Trans. Japan Soc. Mech. Eng.* **36**, 1117 (1970).
269. Yokogawa, A., and Isaka, M., Experimental investigation of spout diameter in the spouted bed. *Hitachi Zosen Tech. Rev.* **45**, 203 (1970).
270. Yokogawa, A., Ogino, E., and Yoshii, N., Particle velocity in the annulus of the spouted bed. *Trans. Japan Soc. Mech. Eng.* **37**, 1979 (1971).
271. Yokogawa, A., and Isaka, M., Pressure drop and distribution of static pressure in the spouted bed. *Hitachi Zosen Tech. Rev.* **46**, 47 (1971).
272. Yokogawa, A., Ogino, E., and Yoshii, N., Fluid velocity profile in the spout of the spouted bed. *Trans. Japan Soc. Mech. Eng.* **37**, 2135 (1971).
272a. Yokogawa, A., Ogino, E., and Yoshii, N., Flow pattern of fluid in the annulus of the spouted bed. *Trans. Japan Soc. Mech. Eng.* **38**, 148 (1972).

The following papers were presented at the *Joint A.I.Ch.E.-C.S.Ch.E. Conf., 4th, Vancouver, Brit. Columbia, September 9–12, 1973.*

273. Berquin, Y. F., Application of the spouting bed principle to continuous granulation of liquid sulphur on an industrial scale. PEC-Eng., Paris, France.
*274. Brunello, G., Della Nina, G., Nunes, F. C. S., and Nascimento, C. A. O., Minimum air requirement for spouting mixed particles. Univ. of Sao Paulo, Brazil.
*274a. Brunello, G., Peck, R. E., and Della Nina, G., The drying of barley malt in the spouted bed dryer. Univ. of Sao Paulo, Brazil.
275. Dayan, J., Rappaport, D., and Reiss, J., Gas residence time distribution in spouted beds and in fluidized beds. Israel Inst. of Tech., Haifa, Israel.
276. Kmiec, A., Research on heat and mass transfer in spouted beds. Univ. of Toronto, Canada.
*277. Littman, H., Vuković, D. V., Zdanski, F. K., and Grbavčić, Ž. B., Pressure drop and flowrate characteristics of a liquid phase spout-fluid bed at the minimum spout-fluid flowrate. Univ. of Belgrade, Yugoslavia.
*278. Lim, C. J., and Mathur, K. B., Residence time distribution of gas in spouted beds. Univ. of Brit. Columbia, Vancouver, Canada.
*279. McNab, G. S., and Bridgwater, J., The application of soil mechanics to spouted bed design. Univ. of Oxford, England.
280. Mann, U., and Crosby, E. J., Controlled-cycle spouted beds: State of the art. Univ. of Wisconsin, Madison, U.S.A.
*281. Nagarkatti, A., and Chatterjee, A., Pressure and flow characteristics of a gas phase spout-fluid bed. Indian Inst. of Technology, Powai, Bombay, India.
*282. Prakash, C. B., Mathur, K. B., and Murray, F. E., Odour control from a kraft mill lime kiln using a spouted bed pre-drier. Univ. of Brit. Columbia, Vancouver, Canada.
283. Ratcliffe, J. S., Low temperature coal carbonization in spouted beds. Univ. of New South Wales, Kensington, Australia.
*284. van Velzen, D., Flamm, H. J., Langenkamp, H., and Casile, A., Motion of solids in spouted beds. Euratom, Ispra, Italy.
*285. van Velzen, D., Flamm, H. J., and Langenkamp, H., Gas flow pattern in spouted beds. Euratom, Ispra, Italy.
*286. Vuković, D. V., Zdanski, F. K., Vunjak, G. V., Grbavčić, Ž. B., and Littman, H., Pressure drop and liquid holdup in a three phase spouted bed contactor. Univ. of Belgrade, Yugoslavia.

* Published in *Can. J. Chem. Eng.* **52** (April, 1974), with conference discussions.

Subject Index

A

Abbreviations, 286
Adsorption and desorption, 212-213
Aerosol removal, 226
Agglomeration, 227-230, 233, 236, 274
Annulus
  defined, 2
  particle flow lines in, 72-74
  particle motion in and from, 72-85
  particle velocity at wall, 74-75, 79-81, 247, 282
  plasticity and solids flow from, 84-85
  reverse gas flow in, 261-262
  voidage in, 104
Applications, 5, 187-242, *see also* individual operations and processes
  chemical, nonreacting solids, 236-242
  chemical, reacting solids, 227-237
  diffusional, 191-213
  mechanical, 11, 218-226
  summary of, 187-191
  thermal, 214-218
Attrition of solids, 125-130
  applied to comminution, 220-226
  applied to drying on inert solids, 203-205
  in coal carbonization, 230
  in granulation, 202
  in iron ore reduction, 233
  pharmaceutical tablets, 209
  scale-up and, 130, 271

in shale pyrolysis, 232
solids feed arrangement and, 278

B

Bed level control, 278
Bed support, 278-279
Blending of solids, 218-220

C

Cement clinker production, 235-236
Channeling, 12
Charcoal activation, 236
Chemical reactors
  applications of, 226-242
  theoretical analysis for gas phase reaction, 174-186
Coal carbonization, 227-231
  design procedure for, 231
Coating of particles, 188, 208-212
  by evaporation, 208-210
  pyrolitically, 210-212
Comminution, 220-226
Conical vessels, spouting in, 2-3, 5
  for granulation, 201
  maximum bed depth for, 114
  minimum spouting velocity for, 37-38
  object-to-bed heat transfer for, 154
  particle velocity profiles for, 72
  peak pressure drop for, 27-28